RESOURCES FOR TECHNICAL COMMUNICATION

Second Edition

New York Boston San Francisco
London Toronto Sydney Tokyo Singapore Madrid
Mexico City Munich Paris Cape Town Hong Kong Montreal

Senior Vice President and Publisher: Joseph Opiela
Senior Sponsoring Editor: Virginia L. Blanford
Senior Supplements Editor: Donna Campion
Electronic Page Makeup: Lorraine Patsco

Copyright © 2007 Pearson Education, Inc.

For permission to use copyrighted material, grateful acknowledgement is made to the copyright holders on pp. 140–141 whichare hereby made part of this copyright page.

All rights reserved. Printed in the United States of America. Instructors may reproduce portions of this book for classroom useonly. All other reproductions are strictly prohibited without prior permission of the publisher, except in the case of brief quota-tions embodied in critical articles and reviews.

Please visit our website at www.ablongman.com, and find additional resources at www.MyTechCommLab.com.

ISBN: 0-321-45081-7

1 2 3 4 5 6 7 8 9 10 - CW - 09 08 07 06

CONTENTS

To the Instructor   v

Part 1: Sample Documents   1

   Correspondence

      Letters   2

      Memos   16

      E-mails   28

      Career Correspondence   32

   Proposals   42

   Abstracts   76

   Instructions and Procedures   79

   Descriptions and Definitions   82

   Reports   86

   Oral Presentations and PowerPoints   110

Part 2: Scenarios and Case Studies   114

Sources   140

To the Instructor

Resources for Technical Communication was created in response to many requests from technical communication instructors for a larger bank of sample documents to use in their classrooms. Although virtually all technical communication textbooks include sample documents, there never seem to be enough. This book includes samples in all the primary categories—correspondence (including letters, memos, e-mails, and career correspondence and résumés), proposals, abstracts, instructions and procedures, descriptions and definitions, and reports—as well as a slide show presentation. Some of the documents were written by students and some by working professionals, since instructors have requested samples from both contexts.

Part 1: Sample Documents. The introduction to each section of documents provides basic information about that category of documents, as well as a list of quick reminders about purpose, audience, and strategies for producing such documents. The first document in each section is annotated both with labels showing standard formatting features and with rhetorical notes about the purpose of each paragraph or section. Additional selected documents are annotated more extensively, with remarks about format, audience, and style. The remaining documents in each section are not annotated, in order to give instructors the widest possible flexibility in designing assignments for their students.

These documents were selected not because they are perfect models in their categories but because they represent typical written interactions in the education, business, professional and technical worlds. Many have room for improvement. Report #1, for example, reflects the typical structure of a progress report but also includes a number of redundancies; instructors might want to assign students to restructure and shorten this document by one-fourth. Similarly, Proposal #3 could benefit from revision for length. Memo #4 (Announcement) provides ample material for discussion about audience and tone, and Memo #5 (Announcement) is clearly a candidate for an editing assignment.

These documents are provided without suggested activities or exercises so that instructors have the freedom to create their own pedagogical uses for them. There are a myriad of possibilities for use in the classroom and as outside assignments. For example,

- Ask students to study the annotations of a sample document and then provide a similar rhetorical analysis for another in that category.
- Assign students to revise documents for structure, length or use of language.
- Assign students to revise a memo into a letter, or an e-mail into a memo, to reflect the different rhetorical purposes of these kinds of documents.
- Ask students to work with partners, one taking the role of employee and one of manager, and have the employee produce a document similar in purpose to one included here. Then have the manager comment about what works, what doesn't, and how the employee might revise the document.
- Put examples from this collection onto overheads and have students analyze them in class.
- Ask students to identify the differences and similarities in rhetorical strategies between two documents in the same category—the good news and bad news letters, for example, or letters of complaint and of appreciation, or a progress report and a research report.
- Have students create chronological and functional résumés for themselves and discuss which format might work better, given their backgrounds.

Part 2: Case Studies and Scenarios. At the back of this book, you will also find a selection of case studies and brief scenarios. These are intended for classroom use by instructors who wish to incorporate a case study approach in their teaching, although they might also serve as documents for rhetorical analysis, since case studies are sometimes written in business, professional and technical settings for a variety of purposes including training.

Most of the sample documents in this collection and all of the case studies and scenarios are drawn from technical communication textbooks published by Pearson Longman. Source information for the content in this book can be found on the last page of the book.

Pearson Longman hopes that you will find these materials useful, and we welcome your comments on this resource. For a wide range of additional resources for your course, visit www.MyTechCommLab.com. Ask your Pearson Longman sales representative for a demonstration access code.

Part 1: Sample Documents

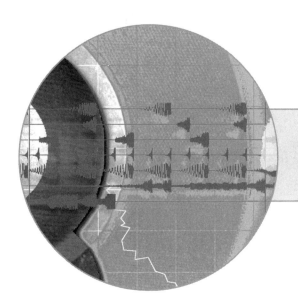

Letters

Regardless of their content, letters are a kind of gift. They cost money (letterhead and postage), and they take more time than e-mail to produce and send. They provide a more formal context for correspondence than e-mails, and they are therefore an important part of any business's communications.

Letters also provide a paper trail. You should always keep a copy of any letter you receive along with any you send in response. It may seem overly cautious, but it doesn't hurt to scan and store electronically all paper correspondence.

Letters generally serve one of two purposes: an exchange of information (the letter is providing unsolicited information, asking for information or responding to a request for information); or conveyance of a complaint (the letter is communicating a complaint or responding to one). Letters can also simply accompany larger documents; these are called transmittal or cover letters.

Letters follow a typical pattern:
- address and salutation,
- background or context,
- specific instructions about how to respond to the letter,
- thank you,
- signature.

But each specific type of letter often has its own form: for example, good-news letters put the primary information at the beginning; bad-news letters often place it at the end. It makes sense to keep a file of sample letters of all kinds. Your workplace may in fact have a collection of "boilerplate" or template letters that you should use whenever possible.

Quick Reminders for Writing Letters

- **Consider your audience.** Most letters go to people outside your company and often to people you don't know. Write any letter as if the person you are writing to is standing directly in front of you. What frame of mind is your recipient likely to be in?

How can your letter address that state of mind?

- **Be brief, direct, and clear.** The recipient is no doubt as busy as you are. Explain clearly the intent of the letter, and tell the recipient exactly what he or she needs to know and do in response.

- **Be correct and credible.** Make sure your spelling and grammar are correct. Check details before you put a letter in the envelope: the date, the recipient's name and address, your name, your return address, your signature. If you have mass produced the letter, as with a letter inquiring about a job, be absolutely certain you've got the right letter in the right envelope and that you've signed each one.

Letter 1: Good News

Heading

Parametrics, Inc.
350 South Street
Pittsburgh, PA 15213-1821

Date

15 January 2006

Internal address

Terry Miller
238 Chestnut Street, #3B
Butler, PA 16001-0238

Salutation

Dear Ms. Miller:

Presents good news first

It gives me great pleasure to confirm our verbal offer and your acceptance to join Parametrics, Inc., as a technical documentation specialist in the engineering division.

Describes contractual details

Your compensation will include your monthly salary of $XXXX (which is equivalent to $XXXXX annually) plus the benefits outlined in the enclosed summary. After three months of probation, you will be eligible to receive fourteen days vacation during 2000. According to the terms of our current policy, your performance and salary will be reviewed in 2001. Our regular working hours are from 9:00 a.m. to 5:00 p.m., Monday through Friday.

Provides instructions

On your first day, please report directly to the Human Resources Department at 9:00 a.m. to be entered onto payroll, arrange orientation, and initiate the administrative procedures. As part of these procedures, and as a matter of compliance with federal law, we require you to bring proof of your eligibility to work at that time. Please refer to the enclosed pink sheet for specifics regarding which credentials to bring. If you have any questions, please give me a call.

Expresses confidence in employee

We believe that you will make a significant contribution to Parametrics, and at the same time will realize the professional growth you seek.

Requests response

As soon as possible, please acknowledge your acceptance of this job offer by signing the enclosed copy of this letter and returning it to me. We very much look forward to your joining the company on February 1.

Closing

Best regards,   Accepted by:

Signature

Ross Wagner
_____

Ross Wagner
Senior Employee Relations Representative

Enclosures

cc: Human Resources Files

Letter 2: Bad News

June 19, 2005

Ms. Diane Russell
3923 Fourth Ave.
Detroit, Michigan 94746

Dear Ms. Russell:

When we here at FORM-Tech decided to hire a new Sales Operation Manager, we were not prepared for the number of highly qualified individuals who would apply. After reviewing over sixty applications, we narrowed the field to five exceptional candidates. You were one of those candidates because of your fine credentials and excellent work record.

Because we are able to interview only two of the remaining five applicants, we had the difficult job of narrowing the field one more time. We carefully reviewed each candidate's education, work experience, and special skills. Each potentially brought varied and worthwhile qualities to our company. We had to find some way to narrow an impressive field.

We decided, therefore, to select those applicants who had specific job-related experience in field management, because the new Sales Operation Manager at FORM-Tech will spend a good deal of time working with our representatives at satellite offices. As a result, we must regretfully inform you that we will not be able to offer you the position.

We commend you on an excellent resume and dossier and wish you continued success in the future.

Thank you very much for your interest in FORM-Tech.

Sincerely,

Margaret Findlay
Vice President, FORM-Tech
8582 W. Front Street
Lansing, Michigan 94722

Letter 3: Claim

This writer's task is to convince the reader to replace or repair a product, so this might be considered a "bad news" letter for the manufacturer. However, the manufacturer has the power to deny the claim, so the writer must figure out how to get satisfaction from a potentially unwilling manufacturer. Every other choice—including organization, style, and design—should be dictated by this rhetorical situation and purpose.

[Par 2 "Here is what happened..."]
Here the writer tells the reader clearly and directly what happened—a necessary strategy, since the manufacturer will not replace or repair the camcorder without knowing this information.

[Par 3 "We paid a significant amount..."]
This sentence refers directly to the manufacturers' advertising, implicitly challenging the manufacturer to honor its claims. Would an explicit challenge work equally well? How might your reader react if you wrote: "You claim that your camera is 'highly durable,' but apparently it cannot live up to that standard."

[Par 4 "Please repair..."]
Here the writer makes a direct request of the manufacturer. If there's something you want your audience to do, state it clearly. If your letter simply makes a complaint, the recipient can easily ignore it.

[Par 5 "Thank you..."]
Always provide contact information. You don't want to make it difficult to respond.

[Par 5]
Saying "please" matters. Notice also that the writer "requests" rather than "demands" (Paragraph 1); the writer is aware of potential reader reaction.

Outwest Engineering

2931 Mission Drive, Provo, UT 84601 (801) 555-6650

June 15, 2004

Customer Service
Optima Camera Manufacturers, Inc.
Chicago, IL 60018

Dear Customer Service:

We are requesting the repair or replacement of a damaged ClearCam Digital Camcorder (#289PTDi), which we bought directly from Optima Camera Manufacturers in May 2004.

Here is what happened. On June 12, we were making a promotional film about one of our new products for our website. As we were making adjustments to the lighting on the set, the camcorder was bumped and it fell ten feet to the floor. Afterward, it would not work, forcing us to cancel the filming, causing us a few days' delay.

We paid a significant amount of money for this camcorder because your advertising claims it is "highly durable." So, we were surprised and disappointed when the camcorder could not survive a routine fall.

Please repair or replace the enclosed camcorder as soon as possible. I have provided a copy of the receipt for your records.

Thank you for your prompt response to this situation. If you have any questions, please call me at 801-555-6650, ext. 139.

Sincerely,

Paul Williams

Paul Williams
Senior Product Engineer

Letter 4: Adjustment

The author has decided in advance on the rhetorical purposes: to apologize and adjust; therefore, the introduction expresses regret and states the plan of action. The author has also identified the recipient as outside the company, making a letter the appropriate genre, as opposed to a memo.

[Par 1]
The author uses "we" in reference to the company in an attempt to maintain goodwill with the customer and immediately expresses regret without accepting blame.

[Par 2]
The body paragraphs clearly describe what actions have been taken and what future actions will occur; they describe a plan of action for the customer. Moreover, the author shows appreciation to the customer and names the specific product in question.

[Concluding Par]
Any essential information (aside from contact phone number) should be introduced before the conclusion. This conclusion is successful because of its brevity, lack of repetition, and lack of new essential information.

O C M
Optima Camera Manufacturers, Inc.
Chicago, IL 60018   312-555-9120

July 1, 2004

Paul Williams, Senior Product Engineer
Outwest Engineering Services
2931 Mission Drive
Provo, UT 84601

Dear Mr. Williams,

We are sorry that the ClearCam Digital Camcorder did not meet your expectations for durability. At Optima, we take great pride in offering high-quality, durable cameras that our customers can rely on. We will make the repairs you requested.

After inspecting your camera, our service department estimates the repair will take two weeks. When the camera is repaired, we will return it to you by overnight freight. The repair will be made at no cost to you.

We appreciate your purchase of a ClearCam Digital Camcorder, and we are eager to restore your trust in our products.

Thank you for your letter. If you have any questions, please contact me at 312-555-9128.

Sincerely,

Ginger Faust

Ginger Faust
Customer Service Technician

Letter 5: Response to Claim

July 18, 2005

Mr. Charles Harrison
4876 Bluff Street
Boulder, Colorado 29127

Dear Mr. Harrison:

We here at ISIX, Inc. pride ourselves not only on the quality of our software products but also in the faith our repeat customers have in us. As a result, we are happy to refund, in full, the $453.95 that you spent on our DesignDraw program in March 2005. We are sorry that the program never worked properly for you and can only guess there was a flaw in manufacturing.

We have received no other claims for refund on this particular product, so we hope that yours was an isolated situation. In addition to the full refund, we are sending you a $50.00 gift certificate for use toward any other ISIX product.

Our customers are important to us, so we will carefully examine the damaged product you have returned. We apologize for any inconvenience you have experienced as a result of this situation and hope you will consider purchasing a replacement DesignDraw program, one of our most popular products.

Sincerely,

Rick Aston
Vice President, ISIX Inc.
4716 South Summit St.
Oakland, CA 95736

Letter 6: Collection

Greene's

New Acres Mall   Tallahassee, FL 32301

June 16, 2004

Mr. William Britton
55-A Jackson Road
Tallahassee, FL 32301

Dear Mr. Britton:

We appreciate your continued patronage of Greene's. We note, however, that your charge account is now $565.31 overdue, and that we have not received your monthly payment since April.

If you have recently sent in your payment, please ignore this friendly reminder. If not, we would appreciate a minimum remittance of $50.00
at your earliest convenience.

If you have any questions about your account, please call us at 555-0123, Ext. 123.

Sincerely,

Heather Sutcliffe

Heather Sutcliffe
Credit Services Department

Letter 7: Complaint

2708 Bryonhall Drive
Austin, TX 78745
December 7, 2005

Mr. Larry Watkins, Manager
Mirabelle's Department Store
165 E. Main Street
Austin, TX 78701

DRESS WITH DEFECTIVE ZIPPER

Dear Mr. Watkins:

On December 3, 2005, I purchased a party dress for my daughter from your selection of Elegant Christmas Frocks displayed in your preteen department. I was assisted by Aileen Dawes in making my selection. Since my daughter was not with me, I asked Ms. Dawes if it would be possible to return the dress if it did not fit her or if she did not like it. Ms. Dawes assured me there would be no problem.

I took the dress home and had my daughter try it on that evening. As I zipped it up, the zipper came loose from the dress. I inspected the zipper and noted that the seam to which the zipper should have been sewn was too narrow and was frayed in parts.

On December 5, I returned the dress to your store. Ms. Dawes was not working that day, but another salesperson, Christine Mays, told me that I could not return the dress because it was damaged. She referred me to her supervisor, Mary Kaiser, when I protested that the dress was defective. Ms. Kaiser said that store policy prohibited the return of damaged goods.

I would like to protest this decision. I have been a Mirabelle's customer for many years and have come to associate your store with exceptional quality and outstanding service. This is the first item I have purchased that did not meet the standards I expect from Mirabelle's. Obviously the garment was not properly sewn. It is an expensive garment, and all I am asking is to have it replaced with one that is not defective.

I'm sure that the Mirabelle's return policy for damaged goods was not meant to include items that are defective. I would like to discuss this matter with you at your convenience in order to work out a fair adjustment. My telephone number is 555-0467; please call me with a time that I can meet with you.

Sincerely,

Kathy Judge

Kathy Judge

Letter 8: Complaint

1201 Winterwood Boulevard
Oakland, CA 94601
June 25, 2004

Ms. Phyllis A. Dow
Bay Professional Consultants
934 N. Orange Avenue
San Francisco, CA 94103

Dear Ms. Dow:

Over the past few years, you have provided valuable advice to my wife and me concerning our son, Jeff. We appreciate the attention and time you have devoted to our situation. It isn't often that we encounter such professionalism. However, I am writing concerning a billing issue that I hope can be resolved soon.

Recently, we experienced a misunderstanding with your billing office. After our last appointment with you, we stopped by the front desk, where we learned we had a balance due in the amount of $55.00. The staff member did not know what the charge was for and said she would find out on the next day. Because we were in a hurry to pick up our son, I asked my wife to pay the amount, feeling we could sort out the details later.

The next day we found out that the $55.00 is a charge for our supposedly missing an appointment on May 29 at 2:00 p.m. This news was a big surprise to my wife and me. I checked my detailed phone log and verified that I had called your office on the morning of May 27 to cancel the appointment and reschedule it for a later date. In addition to the entry in my log, I specifically recall the telephone conversation.

In sum, my wife and I ask that you and your office staff void the $55.00 charge. It would be unfair to charge us $55.00 when we canceled our May 29 appointment more than 48 hours in advance. I am particularly conscientious about keeping all appointments I make, and if I cannot keep an appointment, I always cancel well before the appointment.

If you would like to discuss this matter with me, please call me at 555-2627.

Considering the amount of time my wife and I have spent with you, this particular incident is relatively minor. We hope it will be resolved amicably, and we look forward to future meetings with you.

Thank you.

Sincerely,

Bob Smith

Bob Smith

Letter 9: Appreciation

BioTechnology, Inc.

2846 Sixth Avenue
Seattle, Washington 98121
Phone: 206-441-8577
Fax: 206-728-4429

July 27, 2005

Dr. Betsy Geist
Laboratory C-2
Building 6

Dear Betsy:

Thank you for your extraordinary service on the CHR4-pill project. Your colleagues have expressed to me their gratitude for your willingness to work overtime to provide them with the essential technical expertise they needed to complete the project on time. Without your expertise and help, we could not have met the necessary deadlines.

As a result of your efforts, BioTechnology was able to submit the CHR4 prototype for beta tests ahead of our nearest competitors. Because of your dedication, BTI remains a leader in pharmaceutical research.

Sincerely,

Paul Smith

Paul W. Smith
Vice President for Research

Letter 10: Appreciation

VALUE-RITE OFFICE SUPPLIES

462 Decatur Street • Atlanta, GA 30300 • (404) 555-1234

March 19, 2006

Ms. Helen Reynard, Owner
Reynard's Auto Palace
Central Highway
Atlanta, GA 30300

Dear Ms. Reynard:

For the past 10 years, Value-Rite Office Supplies has purchased all our delivery vans from your dealership, and we have relied on your service department for routine maintenance and necessary repairs. During that time I have been repeatedly impressed by the professionalism of your employees, especially Jarel Carter, who staffs the service desk.

Both in person and on the telephone, Jarel has always been exceptionally knowledgeable, helpful, and courteous and is always willing to go the extra mile to ensure customer satisfaction. Just last week, for example, he interrupted his lunch break to get me some information about a part that has been on back-order.

If you can continue to attract employees of Jarel's caliber, you shouldn't have any difficulty remaining the area's #1 dealership. Be sure to keep him in mind the next time you're considering merit raises!

Sincerely,

Gary Richie

Gary Richie, Owner

Letter 11: Request or Inquiry

Northern Montana Association for the Disabled
1010 South First Street
Big Valley, Montana 59822

The Sunset Inn
1010 Lake Drive
Big Valley, Montana 59822
June 1, 2005

Dear Manager/Owner:

It has come to our attention that your inn does not contain a room designed for guests requiring wheelchairs. The technology and fixtures to accommodate *all* guests is affordable and readily available. We ask that you consider renovating one of your first-floor rooms with a special shower and lower counters, beds, shelves, dresser, night stands, and door locks.

While you are not required, under the Americans with Disabilities Act, to provide such accommodations (because your building is an older structure), our organization will list your inn in our yearly recreation guide for disabled people if you do. We hope you believe, as we do, that everyone should be able to travel and find comfortable lodging in every town in the our region. We are counting on you to provide such comforts to the disabled, should you make the suggested changes.

Sincerely,

J. Swanson

Jessica Swanson, Chair
Northern Montana Association for the Disabled

Letter 12: Request or Inquiry

The Weekly News

P.O. Box 123
Littleton, New York 13300
Telephone (315) 555-1234 • Fax (315) 555-4321

February 24, 2006

Chief Joseph Kealy
Littleton Police Department
911 Main Street
Littleton, NY 13300

Dear Chief Kealy:

    It is our understanding that a Littleton resident, Mr. Alex Booth, is the subject of an investigation by your department, with the assistance of the county district attorney. In keeping with the provisions of the New York Freedom of Information Law, I am requesting information about Mr. Booth's arrest.

    This information is needed to provide our readership with accurate news coverage of the events leading to Mr. Booth's current situation. The Weekly News prides itself on fair, accurate, and objective reporting, and we are counting on your assistance as we seek to uphold that tradition.

    Since the police blotter is by law a matter of public record, we will appreciate your full cooperation.

Sincerely,

Nancy Muller

Nancy Muller, Reporter

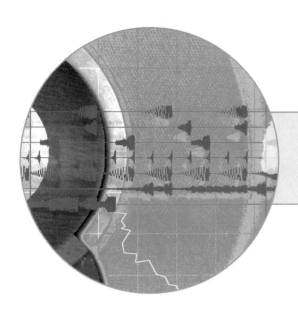

Memos

Memoranda are internal letters. They provide information for people within the same organization. Because they are internal, they often do not need to include the level of explanation and detail that you might want to include in a letter. Organizations use memos to make announcements, disseminate information, or confirm understandings. Memos often provide initial agendas for meetings and summaries after meetings conclude. Memos are less formal than letters, but they also provide a paper trail for organizational process; keep in mind that your memos will form part of the official office files, so they should be written with care.

The general pattern of a memo includes: a company and/or department logo; *To, From, Date* and *Subject* lines; context or background; specific request or announcement; and a clear indication of what response is needed. Although the samples included here are all one-page memos, many longer documents are formatted in this way. See, for example, Report #1 (Progress Report) and Proposal #3.

Quick Reminders for Writing Memos

- **Be direct and professional.** Unlike a letter, memos do not need pleasantries or a formal conclusion.
- **Emphasize the important points.** Use bullets or numbered lists to identify main points when appropriate, rather than narrative.
- **Know your audience.** Put yourself in the shoes of a naïve reader: If you didn't know very much about the subject of this memo, would you understand it?

Memo 1: Transmittal or Cover Memo

Name and title of recipient	**TO:** Priscilla Walton, Director of Production
Name and title of writer	**FROM:** Sally Shore, Editor **CC:** Joe Martinez, Editor in Chief
Date provides chronological record	**DATE:** 12 May 2004
Clear subject line	**RE:** **Release to Production:** Charmin, *The Comprehensive Handbook* 3E

Herewith the initial materials for this text. Attached please find:

List of documents included
- Original and 3 copies of ms. for **Chs. 1–18** (of 53).
- 3 copies of the current edition
- Image log and image manuscript for Chs. 1–18.
- Icon ms. for Chs. 1–18.
- Yellow folder including copies of:
  - Author contract
  - PPS printout
  - P&L
  - Manuscript turnover form
  - Cover design spec sheet
  - Permissions initiation form

Indicates special circumstances

Please note: Tearsheet ms. is from *The Comprehensive Handbook*, **Second Edition** with the exception of the insert pp. for ms. p. 9 (pp. 9a–9d); these pages are picked up from the *Comprehensive Handbook Brief Version*, Second Edition, just published.

Still to come:

List of pending documents
- Complete new TOC
- Ms., image log and image ms. for Chs. 19–53. **Due 6/15.**
- Revised Preface. **Due 6/15.**
- Complete marginal icon ms. for Chs. 1–53. **Due 7/15.**
- Front and back cover inside copy. **Due 6/15.**

Memo 2: Announcement

TO: All Travelers and Arrangers
FROM: General Delivery: Travel
DATE: Thursday, June 20, 2005
SUBJECT: Announcing New Travel Website!

Chamberlain Travel Services is pleased to provide a central source for all travel-related information with the release of the new Chamberlain Travel Website. For the first time, Chamberlain employees, travelers and arrangers from all operating groups will have access to general travel information. The Website includes information relating to:
- Chamberlain and Carlson Wagon-Lits program specifics
- Preferred vendor program information
- Chamberlain Hotel Guide
- Key contact information
- Chamberlain travel guidelines
- Corporate Card program details
- Destination information
- Directions from key office locations to major airports
- And much more...!

To access the Chamberlain Travel Website, simply use the Logon ID **chamberlain** and the Password **travel** (lower case) when you click on the designated link: http://chamberlaintravel.com. Please note that the Travel Website contains Chamberlain program information and should be regarded as confidential. Access to the site should be limited to Chamberlain employees. Site information will be updated regularly.

Memo 3: Announcement

In a memo, the writer must first and foremost consider purpose: Why am I writing this memo? What does my audience need to know? But look at this memo. Does the "Re:" line tell readers the purpose of the memo? Read the first seven lines. Do you know now what the memo is about? It has something to do with safety, but what? In fact, this is an urgent memo. But its organization makes that hard to see.

To stand out from the mass of communications we receive every day, memos must be designed to make them easy for the reader to take in. Paragraphs play a role in design; large chunky paragraphs look different from short sleek ones. Which kind of paragraph will be most effective? How else might this writer make the design of this memo work more effectively?

ChemConcepts, LLC

Memorandum

Date: November 14, 2004
To: Laboratory Supervisors
cc: George Castillo, VP of Research and Development
From: Vicki Hampton, Safety Task Force
Re: FYI

It is the policy of the ChemConcepts to ensure the safety of its employees at all times. We are obligated to adhere to the policies of the State of Illinois Fire and Life Safety Codes as adopted by the Illinois State Fire Marshal's Office (ISFMO). The intent of these policies is to foster safe practices and work habits throughout companies in Illinois, thus reducing the risk of fire and the severity of fire if one should occur. The importance of chemical safety at our company does not need to be stated. Last year, we had four incidents of accidental chemical combustion in our laboratories. We needed to send three employees to the hospital due to the accidental combustion of chemicals stored or used in our laboratories. The injuries were minor and these employees have recovered; but without clear policies it is only a matter of time before a major accident occurs. If such an accident happens, we want to feel assured that all precautions were taken to avoid it, and that its effects were minimized through proper procedures to handle the situation.

In the laboratories of ChemConcepts, our employees work with various chemical compounds that are can cause fire or explosions if mishandled. For example, when stored near reducing materials, oxidizing agents such as peroxides, hydroperoxides and peroxyesters can react at ambient temperatures. These unstable oxidizing agents may initiate or promote combustion in materials around them. Of special concern are organic peroxides, the most hazardous chemicals handled in our laboratories. These compounds form extremely dangerous peroxides that can be highly combustible. We need to have clear policies that describe how these

Memo 3: Announcement (continued)

Finally, we reach the crux of this message—a mandatory meeting, and the instructions for preparing for it. How could the writer organize this memo more effectively?

kinds of chemicals should be stored and handled. We need policies regarding other chemicals, too. The problem in the past is that we have not had a consistent, comprehensive safety policy for storing and handling chemicals in our laboratories. The reasons for the lack of such a comprehensive policy are not clear. In the past, laboratories have been asked to develop their own policies, but our review of laboratory safety procedures shows that only four of our nine laboratories have written safety policies that specifically address chemicals. It is clear that we need a consistent safety policy that governs storage and handling of chemicals at all of our laboratories.

So, at a meeting on November 1, it was decided that ChemConcepts needs a consistent policy regarding the handling of chemical compounds, especially ones that are flammable or prone to combustion. Such a policy would describe in depth how chemicals should be stored and handled in the company's laboratories. It should also describe procedures for handling any hazardous spills, fires, or other emergencies due to chemicals. We are calling a mandatory meeting for November 30 from 1:00-5:00 in which issues of chemical safety will be discussed. The meeting will be attended by the various safety officers in the company, as well as George Castillo, VP of Research and Development. Before the meeting, please develop a draft policy for chemical safety for your laboratory. Make fifteen copies of your draft policy for distribution to others in the meeting. We will go over the policies from each laboratory, looking for consistencies. Then, merging these policies, we will draft a comprehensive policy that will be applicable throughout the corporation.

Memo 4: Announcement

TO: All Support Staff
FROM: Hevard Johnson
DATE: November 6, 2005
SUBJECT: Miscellaneous gripes and complaints

    The following is a list of friendly reminders of things that either your mamas didn't teach you or that you forgot at some point in your checkered past. Most of these gentle prods toward respectable behavior apply to everyone. Others, however, may not apply to our subcontractors. I do, however, request that each of you take a moment to familiarize yourselves with the contents of this list and heed the messages found therein.

    • Time sheets are due on either my or Dean's desk by **9:00 a.m.** on the 15th of each month (or the last working day before the 15th) and on the last working day of each month. I am extremely tired of having to go around searching, begging, and pleading for time sheets. You also signed an oath during orientation that you would keep these time sheets filled out on a daily basis. I think many of you lied.

    • Our office opens at **8:00 a.m.** I have checked and I am quite certain that nobody changed that to 8:15.

    • **Attention all smokers!!** We now have ash trays mounted beside each of the four entrances to the building. Use them! I'm really, really sick and tired of picking your butts up in the parking lot! You have to walk right past the new ash trays to get back into the building after you have thrown your butt on the ground. Have some pride and common courtesy.

    • The "aluminum" recycling bin in the breakroom is for **aluminum.** What a shock! How could I expect anybody to figure that one out without written directions. There are trash cans for the other trash. Tours can be arranged if you are still having difficulty with this concept.

    • As you are casually staggering down the hallways, up the stairs, or otherwise meandering through the buildings, put your pens or pencils in your pockets, behind your ears, or in some other convenient receptacle. Apparently it was only a pipe dream that you would all learn to walk straight enough to not make huge black marks on the walls. I guess that I will have to use our bonus money to repaint, or make you promise to use it on grace and poise lessons.

    • Eating at your desks is now forbidden. Hansel and Gretel don't work here, so there is no need for the continuous food trails leading through the building. All eating done within the confines of 939 North Washington Street shall henceforth be conducted solely in the break room. Yes, the training room is off limits too!

    • For your own safety—DO NOT stand anywhere near the back door at 4:55. You could be killed in the stampede of people leaving early. I find it strange that people's watches are so cheap that they can gain ten minutes in one day. The same watch can be five minutes late in the morning, yet be five minutes early in the afternoon.

    This list is just the tip of the iceberg. We all need to exercise a little common sense and a little common courtesy rather than just being common. Think about the other people who work here and respect them and their property. We are all responsible adults here; let's behave like it.

Memo 5: Announcement

Memorial Hospital

MEMORANDUM

DATE: September 8, 2003

TO: All Employes

FROM: Roger Sammon, Clerk
          Medical Recrods Department

SUBJECT: Patricia Klosek

As many of you allready know. Patricia Klosik from the Medical records Depratment is retiring next month. After more then thirty years of faithfull service to Memorial hospital.

A party is being planed in her honor. It will be at seven oclock on friday October 24 at big Joes Resturant tickets are $30 per person whitch includes a buffay diner and a donation toward a gift.

If you plan to atend please let me no by the end of this week try to get you're check to me by Oct 10

Memo 6: Request or Inquiry

MEMORANDUM

TO: All Employees

FROM: Alice Jones, Personnel Supervisor *AJ*

DATE: February 15, 2006

SUBJECT: Updated Resumes Required by March 1

Many of you may be aware of the contract ONEX Incorporated recently won with the Department of Natural Resources. The proposed project, a complete survey of 200 acres of government-owned forest in Alaska leading to a proposal for a new national park site, will take a collaborative effort. Company President Harlan Riggs has asked me to collect from all of you updated resumes for our files.

Updating Personnel Files

We are updating company files for several reasons. First, one of the conditions in awarding us this lucrative contract was an updated resume of all employees involved in the project. Second, we need to evaluate both your college-level work experience as well as your subsequent work-related expertise before we assign specific jobs within the project. Third, resumes have not been updated for three years; this seems like a good time to take care of this.

Guidelines for Updating Personnel Files

If you do not have a copy of your old resume, please see me for a file copy. Once you have the old copy, please update the following areas:

- work-related expertise (jobs or special projects you have completed since the last update)
- education (coursework, certificates, or degree programs)
- awards (grants, certificates, or any other accolade)
- special skills (any new and unique skills you have acquired in the last five years)

In addition, please prepare your resume using Word. Submit to me a hard copy with disk no later than March 1, 2006.

We appreciate your cooperation on this project.

Memo 7: Recommendation

Ethical, Legal, and Interpersonal Considerations

Is the information specific, accurate, and unambiguous?
Is this the best report medium (paper, email, telephone call) for the situation?
Is the memo inoffensive to all parties?

Organization

Is the material "chunked" into easily digestible parts?

Format

Does the memo have a complete heading?
Does the subject line announce the memo's contents?
Are paragraphs single-spaced within and double-spaced between?
Do the headings announce subtopics?
Does the document's appearance create a favorable impression?

Trans Globe Airlines

MEMORANDUM
To: R. Ames, Vice President, Personnel
From: B. Doakes, Health and Safety
Date: August 15, 20xx
Subject: **Recommendations for Reducing Agents' Discomfort**

In our July 20 staff meeting, we discussed physical discomfort among reservation and booking agents, who spend eight hours daily at automated workstations. Our agents complain of headaches, eyestrain and irritation, blurred or double vision, backaches, and stiff joints. This report outlines the apparent causes and recommends ways of reducing discomfort.

Causes of Agents' Discomfort

For the time being, I have ruled out the computer display screens as a cause of headaches and eye problems for the following reasons:

1. Our new display screens have excellent contrast and no flicker.
2. Research findings about the effects of low-level radiation from computer screens are inconclusive.

The headaches and eye problems seem to be caused by the excessive glare on display screens from background lighting.
   Other discomforts, such as backaches and stiffness, apparently result from the agents' sitting in one position for up to two hours between breaks.

Recommended Changes

We can eliminate much discomfort by improving background lighting, workstation conditions, and work routines and habits.

Background Lighting. To reduce the glare on display screens, these are recommended changes in background lighting:

1. Decrease all overhead lighting by installing lower-wattage bulbs.
2. Keep all curtains and adjustable blinds on the south and west windows at least half-drawn, to block direct sunlight.
3. Install shades to direct the overhead light straight downward, so that it is not reflected by the screens.

Workstation Conditions. These are recommended changes in the workstations:

1. Reposition all screens so light sources are neither at front nor back.
2. Wash the surface of each screen weekly.
3. Adjust each screen so the top is slightly below the operator's eye level.
4. Adjust all keyboards so they are 27 inches from the floor.
5. Replace all fixed chairs with adjustable, armless, secretarial chairs.

Memo 7: Recommendation (continued)

Content
Is the message brief and to the point?
Are recipients given enough information for an *informed* decision?
Are the conclusions and recommendations clear?

Style
Is the writing clear, concise, exact, and fluent?
Is the tone appropriate?
Does the memo appear to have been carefully proofread?

Work Routines and Habits. These are recommended changes in agents' work routines and habits:

1. Allow frequent rest periods (10 minutes each hour instead of 30 minutes twice daily).
2. Provide yearly eye exams for all terminal operators, as part of our routine healthcare program.
3. Train employees to adjust screen contrast and brightness whenever the background lighting changes.
4. Offer workshops on improving posture.

These changes will give us time to consider more complex options such as installing hoods and antiglare filters on terminal screens, replacing fluorescent lighting with incandescent, covering surfaces with nonglare paint, or other disruptive procedures.

cc. J. Bush, Medical Director
    M. White, Manager of Physical Plant

Memo 8: Internal Information

This memo was intended for a limited, internal audience, but was leaked to the press where it reached a much wider audience. When Secretary Rumsfeld asked "Are we winning or losing the Global War on Terror?" it was intended as a "thought question" to his staff. But when the memo became public, the wider audience interpreted it as a possible admission of failure. Certainly, Secretary Rumsfeld did not intend his memo to reach the public. However, writers should always consider the possibility that unintended readers will see what they write.

October 16, 2003

TO: Gen. Dick Myers
            Paul Wolfowitz
            Gen. Pete Pace
            Doug Feith

FROM: Donald Rumsfeld

SUBJECT: Global War on Terrorism

The questions I posed to combatant commanders this week were: Are we winning or losing the Global War on Terror? Is DoD changing fast enough to deal with the new 21st century security environment? Can a big institution change fast enough? Is the USG changing fast enough?

DoD has been organized, trained and equipped to fight big armies, navies and air forces. It is not possible to change DoD fast enough to successfully fight the global war on terror; an alternative might be to try to fashion a new institution, either within DoD or elsewhere—one that seamlessly focuses the capabilities of several departments and agencies on this key problem.

With respect to global terrorism, the record since September 11th seems to be:

— We are having mixed results with Al Qaida, although we have put considerable pressure on them—nonetheless, a great many remain at large.

— USG has made reasonable progress in capturing or killing the top 55 Iraqis.

— USG has made somewhat slower progress tracking down the Taliban—Omar, Hekmatyar, etc.

— With respect to the Ansar Al-Islam, we are just getting started.

Have we fashioned the right mix of rewards, amnesty, protection and confidence in the US?

Does DoD need to think through new ways to organize, train, equip and focus to deal with the global war on terror?

We all receive hundreds of communications each week in the workplace. To grab an audience's attention, writers must organize documents as effectively as possible. Notice how this memo is organized simply and effectively. The writer states subject and purpose; next, he outlines the current situation with bullet points, making it easy to scan. Given his audience, he can count on their knowing the general situation, so he only has to give brief reminders.

Bullet points can sometimes do more harm than good. The writer has to make sure that bullet point summaries are precise so that key information is not left out.

Memo 8: Internal Information (continued)

At the end of the memo, the writer summarizes key questions. Bullet points help make the information scannable, and the writer uses a different bullet symbol to distinguish the lists.

Finally, the writer states precisely what he wants the audience to do. Notice that the writing style is direct. The reader does not need to be an expert on defense or terrorism to understand the questions and concerns.

Are the changes we have and are making too modest and incremental? My impression is that we have not yet made truly bold moves, although we have made many sensible, logical moves in the right direction, but are they enough?

Today, we lack metrics to know if we are winning or losing the global war on terror. Are we capturing, killing or deterring and dissuading more terrorists every day than the madrassas and the radical clerics are recruiting, training and deploying against us?

Does the US need to fashion a broad, integrated plan to stop the next generation of terrorists? The US is putting relatively little effort into a long-range plan, but we are putting a great deal of effort into trying to stop terrorists. The cost-benefit ratio is against us! Our cost is billions against the terrorists' costs of millions.

- Do we need a new organization?
- How do we stop those who are financing the radical madrassa schools?
- Is our current situation such that "the harder we work, the behinder we get"?

It is pretty clear that the coalition can win in Afghanistan and Iraq in one way or another, but it will be a long, hard slog.

Does CIA need a new finding?

Should we create a private foundation to entice radical madrassas to a more moderate course?

What else should we be considering?

Please be prepared to discuss this at our meeting on Saturday or Monday.

Thanks.

DHR:dh
101503-58

Please Respond by _____

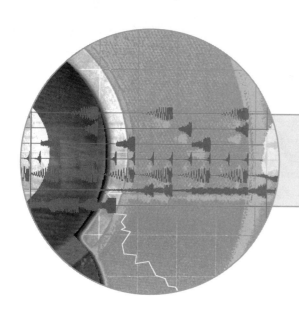

E-mails

Unlike a letter, an e-mail is not a gift. In fact, it is often seen as an imposition on a person's time. Many business people get 100 or more e-mails every day, many of which are without value. "Spam," or unwanted, widely broadcast e-mails, is often filtered out by software maintained on the servers of many big businesses. Nevertheless, many spam messages make their way through the lines of defense. And even the information in many apparently business-related messages could be handled in another, less intrusive way. You should understand that e-mail is an effective way of communicating only if you use it sparingly and wisely.

Quick Reminders for Writing E-mails

- **Use the subject line appropriately.** Be sure that the subject line tells your recipient exactly what the content is. "Stuff" is rarely a good choice.
- **Be brief, clear, and direct.** If you find that you are writing more than a few sentences, consider picking up the phone.
- **Beware of e-mail trails.** Don't simply forward a long trail of e-mails to someone else, expecting the recipient to read through the whole string to find the relevant information. Delete unnecessary information. Make sure that only people who really need to know are copied.
- **Be correct and credible.** Despite the informality of this form, use appropriate and businesslike grammar, spelling, and forms of address. Always reread e-mails before sending them, to make sure that your message is accurate.
- **Read before you send.** Before you hit send, make sure the e-mail address you are using is the one you want to use. Make sure that nothing offensive or private is buried in a string of e-mails. Make sure you are saying what you want to say.

E-mail 1: Request or Inquiry

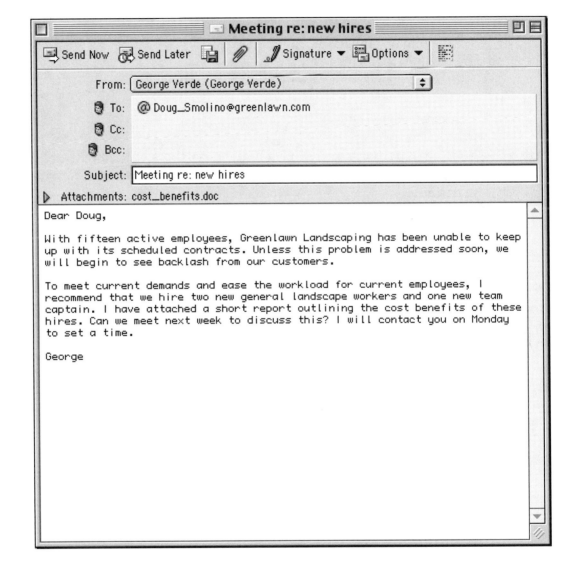

Provides clear statement of subject

States issue succinctly

Proposes solution and options

E-mail 2: Request or Inquiry

From: Blanford, Ginny [mailto:Ginny.Blanford@ABLongman.com]
Sent: Monday, January 26, 2004 8:34 AM
To: 'kwunderlich@utsa.edu'
Subject: FW: Reviewing McMillan

Dear Professor Wunderlich:

> I'm an editor with Allyn & Bacon, the publishers of McMillan and
> Schumacher's Research in Education. We are about to revise this text, and
> I am looking for instructors in educational research to provide feedback
> on the current edition and help us identify what needs to be revised.
> Would you be interested in helping us out in this way? If so, I would
> first need confirmation from you that you do indeed teach a course
> appropriate for this text (as well as information about what book you
> currently use). Then I will send you guidelines for the review. We would
> need your comments back by February 20th, and I can offer you an
> honorarium of $xx for your review.
>
> I hope to hear from you! Many thanks in advance for your time. Please
> let me know if you have questions or need further information before
> making a decision about this project.
>
> Sincerely,
> Ginny Blanford
>
>
>
> Virginia L. Blanford, Ph.D.
> Senior Sponsoring Editor
> Allyn & Bacon/Longman
> 1185 Avenue of the Americas
> New York, NY 10036
> 212-XXX-3372
>   Ginny.Blanford@ablongman.com

E-mail 3: Request or Inquiry

Dear Reviewer,

I am in the process of creating the blogroll for the collection (or, rather, linkroll, since not all the links will be to weblogs). Please send me the URL of the site you'd like to represent you by 7 May; weblogs or homepages are fine, but only one per reviewer. If you did a collaborative review, we would like a link for each of the reviewers.

Thank you,

Clancy

--
Clancy Ratliff
Graduate Instructor, University of Minnesota
http://umn.edu/home/ratli008 | http://kairosnews.org
http://culturecat.net | http://rhetcomp.com

E-mail 4: Announcement

Greetings Everyone,

Just a reminder of the General Education Subcommittee meeting tomorrow (Friday, April 30) at 10 AM in the Center for Teaching and Learning (Room 205 Library South). The Center is located just beyond the circulation desk in Library South. Remember, that you will need your PantherCard for access to the University Library.

See you tomorrow,
Harry

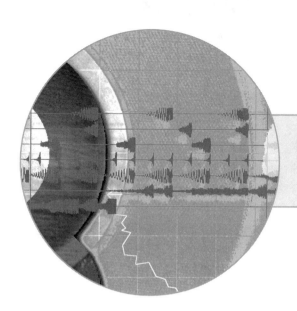

Career Correspondence

Looking for work is an ongoing process and your application materials are a living record. You should keep files on your accomplishments and learn from any letters of rejection you might encounter along the way. You should also view your résumé as a roadmap to the future: Where are you now, where do you want to be, and how are you going to get there?

Your **application letter** is one of the most important documents you will ever create. The letter should prove that you've done your research regarding the company, and it should answer four questions:

- What job are you applying for?

- Where did you see the announcement?

- Why do you want to work for this company? (You need to know more than just whatever is on the company's webpage. Look for books, people who work there already, company publications, news paper articles, anything and everything to demonstrate what you know about this company, and why it appeals to you.)

- And most important, what can you do for them? (Don't just assert what you can do. Offer *evidence* of your having done it in the past, or of your having the skills and commitment to do it in the future.)

As with any letter, proof read your application letter compulsively, and make sure before you stuff it in the envelope that it is properly addressed and signed. You should not use a generic letter, but if the letter is based on a template, make sure you've not included anything that was intended for another company.

Résumés are, first and foremost, a sales document. Yours should be designed to persuade a potential employer that you are the very best person for the job. Your résumé should be organized and formatted to highlight the most important information for the particular position for which you are applying—you may well want to reorganize or reformat it for different kinds of jobs or companies. When you begin the job hunt, you should create two basic résumés—one in plain ASCII code to offer when online submissions are requested, and a traditionally formatted one that can be printed on quality paper for hard copy submissions.

You should also have a portfolio of your work available for submission both electronically and in hard copy, no matter what that work is. If you are looking for work as a technical writer, you should keep an online portfolio of the things you've written.

Your résumé should always include information about your education, your experience, and your skills. If you are just graduating from college, you'll probably want to put your education and perhaps your skills first; if you have more work experience, you will probably want to arrange your résumé chronologically, beginning with the job you are currently holding.

For academic positions, résumés take a slightly different form and are called *curriculum vitae*, or CVs. A CV emphasizes teaching experience and publications and lists awards and fellowships that the writer has won. Also, while a résumé should never exceed two pages—it is designed to provide a summary of career highlights—a CV may be substantially longer and should provide an exhaustive list of publications and accomplishments.

Quick Reminders for Writing Career Correspondence

- **Be correct.** As with all letters—but perhaps even more important here—edit your letter for correct grammar, spelling, and punctuation.
- **Be professional.** Use a friendly but businesslike tone. Avoid exclamation points, slang, and other devices you might use in correspondence with friends.
- **Know your audience.** The goal of most career correspondence—as with most technical writing—is to persuade. In order to persuade your reader to hire you, you need to understand as best you can what benefits you can offer that reader.

Career Letter 1: Application

1766 Wildwood Drive
Chicago, Illinois 60666
July 12, 2006

Mr. Eric Blackmore
Senior Vice President
Alden-Chandler Industries, Inc.
72 Plaza Drive
Milwaukee, WI 53211-2901

Dear Mr. Blackmore:

Provides personal connection

Professor Julia Hedwig suggested that I write to you about an opening for a product chemist in your chemical division. Professor Hedwig was my senior adviser this past year.

Expands on résumé information

I have just completed my B.S. in chemistry at Midwest University with a 3.9 GPA in my major. In addition to chemistry courses, I took three courses in computer applications and developed a computer program on chemical compounds. As a laboratory assistant to Professor Hedwig, I entered and ran the analyses of her research data. My senior project, which I completed under Professor Hedwig's guidance, was an analysis of retardant film products. The project was given the Senior Chemistry Award, granted by a panel of chemistry faculty.

Relates background to specific company

My work experience would be especially appropriate for Alden-Chandler Industries. My internship in my senior year was at Pickett Laboratory, which does extensive analyses for the Lake County Sheriff's office. My work involved writing laboratory reports daily. At both Ryan Laboratories and Century Concrete Corporation, I have worked extensively in compound analysis, and I am familiar with standard test procedures.

Requests response and provides options for additional contact

I would appreciate the opportunity to discuss my qualifications for the position of product chemist. I am available for an interview every afternoon after three o'clock, but I could arrange to drive to Milwaukee any time convenient to you. My telephone number during the day between 10:00 a.m. and 3:00 p.m. is (312) 555-6644.

Sincerely,

Kimberly J. Oliver

Kimberly J. Oliver

Enc.

Career Letter 2: Application

2 January 2004

Dana Lightor
Chair, Department of English
Education University
Anytown, NY 10000

Dear Professor Lightor:

This letter comes in response to your department's advertisement for an instructor in technical and business English, published in the December *Chronicle of Higher Education.* I would like to be considered for that position, based on my experience and skills.

As my enclosed résumé suggests, I have been a professional technical writer for eighteen years, with experience in both corporate and non-profit communications. For the past three years, I have been a freelance writer and have created and managed newsletters for three large companies, as well as handling editing and writing chores for a number of other clients. Perhaps even more important, I have also taught business writing in continuing education programs at our local community college and our two local high schools. I hold a master's degree in English with a specialty in technical communication from your university, and Professor Donaldson of your department can speak to my academic work.

I am enclosing, in addition to my résumé, samples of my work. I hope that you find my qualifications interesting, and I look forward to hearing from you.

Sincerely,

Amanda Bills

Amanda Bills

Career Letter 3: Application

[Par 1]
Human Resources directors may read hundreds of letters for a given position. State your reason for writing *immediately*. You don't need a "catchy" introduction to get the reader's attention. By getting to the point, you're showing your professionalism by respecting the reader's valuable time.

[Par 2]
This writer has clearly done her homework. Presumably, the original advertisement called for someone with a particular kind of education, so this writer highlights that education. The writer also strengthens her credibility by mentioning her mentor and using some technical language she knows her audience will recognize. She comes off like an expert, and that will make her more attractive to this employer.

[Par 3]
Notice that the writer discusses what she would do for the company, *not* what the company might do for her. Too often, writers of application letters discuss what they think the company will do for them: "I hope to get some great experience...." Companies, however, are not interested in doling out experience. They're in business, so you want your letter to tell them how your presence will help their business.

[Par 5]
Conventions are very important in business writing. Readers expect things to be presented a certain way; when they're not, readers can get confused, or even annoyed. It is conventional to close the letter with contact information, so that's precisely what this writer does.

April 2, 2004
834 County Line Rd.
Hollings Point, Illinois 62905

Valerie Sims, Human Resources Manager
Sunny View Organic Products
1523 Cesar Chavez Lane
Sunny View, California 95982

Dear Ms. Sims:

I would like to apply for the Organic Agronomist position you advertised through HotJobs.com on March 19. My experience with organic innovations in plant and soil science as well as my minor in entomology would allow me to make an immediate contribution to your company.

My education and research as an organic agronomist would benefit your company significantly. As a Plant and Soil Science major at Southern Illinois University, I have been studying and researching environmentally safe alternatives to pesticides. Specifically, my mentor, Professor George Roberts, and I have been working on using benevolent insects, like ladybird beetles (*Coleomegilla maculata*), to control common pests on various vegetable plants. We have also developed several varieties of organic insecticidal soaps that handle the occasional insect infestation.

I also worked as an intern for Brighter Days Organic Cooperative, a group of organic farmers, who have an operation similar to Sunny View. From your website, I see that you are currently working toward certification as an organic farm. At Brighter Days, I wrote eleven agronomic plans for farmers who wanted to change to organic methods. My work experience in the organic certification process would be helpful toward earning certification for Sunny View in the shortest amount of time.

Finally, I would bring two other important skills to your company: a background in farming and experience with public speaking. I grew up on a farm near Hollings Point, Illinois. When my father died, my mother and I kept the farm going by learning how to operate machinery, plant the crops, and harvest. We decided to go organic in 1997, because we always suspected that my father's death was due to chemical exposure. Based on our experiences with going organic, I have given numerous public speeches and workshops to the Farm Bureau and Future Farmers of America on organic farming. My farming background and speaking skills would be an asset to your operation.

Thank you for this opportunity to apply for your opening. I look forward to hearing from you about this exciting position. I can be contacted at home (618-555-2993) or through e-mail (afranklin@unsb5.net).

Sincerely,

Anne Franklin

Anne Franklin

(See Anne Franklin's résumé on p. 41)

Career Letter 4: Application

Content
Is the letter addressed to a specifically named person?
Does the letter contain all the standard parts?
Does the letter have all needed specialized parts?
Has the applicant given the recipient all necessary information?
Has the recipient's title been identified?

Format
Does the introduction immediately engage the reader and then lead naturally to the body?
Are transitions between letter parts clear and logical?
Does the conclusion encourage the reader to take action?
Is the format correct?
Is the design appropriate?

Style
Is the letter in conversational language, free of jargon?
Does the letter reflect a "you" perspective throughout?
Does the tone reflect the writer-recipient relationship?
Is the style clear, concise, and fluent throughout?
Is the letter grammatically correct?
Does the letter's appearance enhance the applicant's image?

---

203 Elmwood Avenue
San Jose, CA 10462
April 22, 2006

Sara Costanza
Personnel Director
Liberty International, Inc.
Lansdowne, PA 24153

Dear Ms. Costanza:

Please consider my application for a junior management position at your Lake Geneva resort. I will graduate from San Jose City College on May 30 with an Associate of Arts degree in Hotel/Restaurant Management. Dr. H. V. Garlid, my nutrition professor, described his experience as a consultant for Liberty International and encouraged me to apply.

For two years I worked as a part-time desk clerk, and I am now the desk manager at a 200-unit resort. This experience, combined with earlier customer relations work in a variety of situations, has given me a clear and practical understanding of customers' needs and expectations.

As an amateur chef, I know of the effort, attention, and patience required to prepare fine food. Moreover, my skiing and sailing background might be assets to your resort's recreation program.

I have confidence in my hospitality management skills. My experience and education have prepared me to work well with others and to respond creatively to changes, crises, and added responsibilities.

If my background meets your needs, please phone me any weekday after 4 p.m. at 214–316–2419.

Sincerely ,

James D. Purdy

James D. Purdy

Enclosure

Career Letter 5: Thank You

2705 West Oak Lane
Worcester, MA 01603
August 19, 2003

Mr. Phil Rouse, Manager
Software Documentation Department
Whitcomb Technical Solutions
4703 Haskell Rd.
Bo ton, MA 02122

Dear Mr. Rouse:

I am writing to thank you, Mr. Steve Jenkins, and Mr. Gray Appleton for the time you took out of your busy schedules to interview me this past Monday.

Once again I must say how impressed I am with your staff, your company products, and your company procedures. I am familiar with most of the companies in the Boston area, and your company is easily the most impressive for what it is doing with its software products and documentation.

During our interview, I forgot to tell you that I was under consideration for election to the President's Leadership Council. I just received a letter today notifying me that I have been selected as a member of this council. Every year this council chooses only a handful of students as members, and I am delighted that I have been fortunate enough to be chosen this year.

I am confident I can contribute a great deal to your company, and I hope I have the opportunity to prove myself.

I look forward to hearing from you.

Sincerely

Sarah Chase

Sarah Chase

Résumé 1: Functional

ALICE M. RYDEL
3621 Bailey Drive
Big Rapids, MI 49307
(601) 456-2156

CAPABILITIES

- Use MacIntosh and IBM computers
- Analyze data to prepare organized financial statistics
- Use Lotus and other spreadsheet programs
- Use automated accounting system to produce monthly statements
- Provide software training
- Manage a staff of twelve
- Keep accurate records of large numbers of accounts

ACCOMPLISHMENTS

- Supervised daily data input in a 12,000-customer billing department.
- Set up a simple but efficient file system for record keeping.
- Managed a computer lab available to 200 students.
- Devised a plan to schedule students for maximum lab use.
- Handled inventory of computer lab and submitted requests for materials, equipment, and maintenance.

EDUCATION

2001-2005 John Williams Community College
           AAS Degree, Computer Programming
1997-2001 Murrah High School
           Diploma, college preparatory and basic business courses

WORK EXPERIENCE

2001-2005 John Williams Community College
Part-time Big Rapids, MI
           <u>Computer Lab Assistant.</u> Responsibilities: Supervised 15 labs per week. Scheduled student lab use. Identified needs and submitted requests for materials, equipment, and maintenance

           Midwestern Bell Telephone Company, Billing Department
           Big Rapids, MI
           <u>Night Supervisor.</u> Responsibilities: Supervised night staff. Kept account records.

HONORS AND AWARDS from John Williams Community College

- Data Management Department Award
- Outstanding First-Year Student in Programming
- Citation for excellence in keyboarding skills

LEADERSHIP ACTIVITIES

Phi Theta Kappa Honorary, Projects Committee Chair
College Choir Tour Schedule Committee Chair

Résumé 2: Chronological

Carol Smith

427 Vista Trail; Orlando, FL; (407) 555-2627; csmith@email.com

Objective	Seeking a position as a technical writer or editor.
Education	Bachelor of Arts degree in English. University of Central Florida. May 1987.
Work Experience	
1997 to Present	Senior Technical Writer. *Institute for Simulation and Training.* Orlando, FL. Edit a variety of technical documents including grant proposals and reports.
1995–1997	Senior Technical Writer. *Credit Card Software, Inc.* Orlando, FL. Documented credit card software for programmers and computer operators. Included flowcharts, program narratives, files and records, job steps, and control cards. Used Vollie on the IBM mainframe to scan COBOL source listings and copybooks.
1993–1995	Senior Technical Writer/Trainer. *Valencia Community* College. Orlando, FL. Worked as a writer on a federal grant helping develop training materials for in-house employees for a local software company. Involved producing a taskoriented workbook to enhance and explain how the software worked and how customers used it.
1992–1993	Senior Technical Writer. Software *Design Group.* Orlando, FL. Worked as consultant organizing and revising database administrator's guide.
1990–1992	Senior Technical Writer. *Travelers/EBS, Inc.* Maitland, FL. Documented insurance software to produce user guides, documented hardware, and administered new releases.
1988–1990	Technical Writer. *Dynamic Control Corporation.* Longwood, FL. Documented software for hospital systems. Produced user guides, administrative guides, and reports manuals.
1987–1988	Technical Writer. *Assessment Designs, Inc.* Orlando, FL. Developed exercises for assessment centers, included writing scripts. Role played for assessment of candidates for promotion in major companies.
Computer Skills	Word, PowerPoint, RoboHelp, PageMaker, Excel, HTML.
Honors	Phi Beta Kappa and Cum Laude graduate in 1987.
References	Available on request.

Career Correspondence 41

Résumé 3: Archival

Content
Does the writer include complete contact information?
Does the writer include a brief objective statement?
Does the writer include an education section? A work experience section? A skills section? An awards and membership section?
How does the writer handle references?

Organization/Format
Is the résumé organized into scannable blocks of information?
Does the writer create an effective visual hierarchy of information within each section?
Is the writer's name and contact information prominently displayed?
Is information arranged consistently throughout the résumé?

Style/Design
How does the writer establish emphasis in the résumé?
Are bullets used effectively?
Are items in the writer's lists parallel?
What other design elements does the writer employ?

Anne Franklin
834 County Line Rd.
Hollings Point, Illinois 62905

Home: 618-555-2993
Mobile: 618-555-9167
e-mail: afranklin@unsb5.net

Career Objective
A position as a naturalist, specializing in agronomy, working for a distribution company that specializes in organic foods.

Educational Background
Bachelor of Science, Southern Illinois University, expected May 2005.
 Major: Plant and Soil Science
 Minor: Entomology
 GPA: 3.2/4.0

Work Experience
Intern Agronomist, December 2003–August 2004
Brighter Days Organic Cooperative, Simmerton, Illinois
- Consulted with growers on organic pest control methods. Primary duty was sale of organic crop protection products, crop nutrients, seed, and consulting services.
- Prepared organic agronomic farm plans for growers.
- Provided crop-scouting services to identify weed and insect problems.

Field Technician, August 2003–December 2003
Entomology Department, Southern Illinois University
- Collected and identified insects.
- Developed insect management plans.
- Tested organic and nonorganic pesticides for effectiveness and residuals.

Skills
Computer Experience: Access, Excel, Outlook, PowerPoint, and Word. Global Positioning Systems (GPS). Database Management.
Machinery: Field Tractors, Combines, Straight Trucks, and Bobcats
Communication Skills: Proposal Writing and Review, Public Presentations, Negotiating, Training, Writing Agronomic and Financial Farm Plans.

Awards and Memberships
Awarded "Best Young Innovator" by the Organic Food Society of America
Member of Entomological Society of America

References Available Upon Request

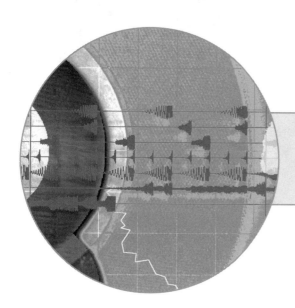

Proposals

A proposal is a document written to persuade some person or group to do something, buy (or fund) something, or change something. As with all persuasive efforts, success depends upon having a clear answer to one question: What does the recipient gain?

Proposals vary widely in length and structure, but typically the parts of a proposal include:

- statement of the problem,
- proposed solution,
- details supporting why the proposed solution is the best (most cost effective, labor efficient, durable, timely, etc.) solution available.

Often you will also have to persuade your audience that you are the best person (or your firm is the best firm) to hire to provide a service, which means providing proof of past successes as well as qualifications.

Included in this section are four related documents: a request for proposals (RFP) addressed to architects for proposals to create a master plan for a small school; responses from two architects; and a proposal from a contractor outlining costs for the work that would be done as a result of the master plan.

Quick Reminders for Writing Proposals

- **Know your audience.** Although this is the first rule for virtually any kind of writing, it is particularly important for proposals because you are trying to persuade your recipient to act.
- **Define the problem.** Before you can convince anyone to do anything, you must prove that a problem or need exists. If your proposal responds to a request for proposals, this is less important—but you still need to demonstrate that you understand the need.

- **Define the solution.** Outline how you will solve the problem or meet the need. Eliminate any information that is not necessary to that purpose—no matter how interesting it may seem to you.

Proposal 1

Cover letter

I. M. Writer
520 Safe Drive St.
Anytown, Anystate, USA 01000
(555) 555-1212

October 13, 2004

Mr. David Decider
Director, Policy Branch
Department of Highways
759 Main Rd. E. Suite 100
Anytown, Anystate, USA 01000

Dear Mr. Decider:

States purpose; describes situation that gave rise to proposal

I am pleased to submit my proposal to study the problem of cyclist head injuries in our state. The economic and personal cost of these injuries is enormous. I believe I can provide the Department of Highways with a thorough and well-thought-out analysis of the problem and present viable solutions to reduce cyclist head injuries.

Defines scope of proposed study

I propose to study solutions to the problem of cyclist head injuries based on statistical and medical research concerning head injuries.

Encourages reader acceptance and follow-up

Thank you for considering this proposal. Please do not hesitate to contact me if you have any questions.

Sincerely,

I. M. Writer
encl.

Provides title, date, and information about writer and recipient, including contact information

A Proposal to Investigate
Bicycle Accident Head Injuries
in Anystate

Oct 13, 2004

Prepared for:
Mr. David Decider
Director, Policy Branch
Department of Highways
759 Main Rd. E. Suite 100
Anytown, Anystate, USA 01000

Submitted by:
I. M. Writer
520 Safe Drive St.
Anytown, Anystate, USA 01000
(555) 555-1212
iwriter@someisp.com

Offers easy access to contents

Table of Contents

Executive Summary	1
An Overview of This Proposal	1
Cycling Head Injuries Are a Severe Problem	1
Some Real-Life Examples of the Problem	2
A Problem for Many	3
What the Study Will Do	4
Possible Solutions to Investigate	5
Action Plan	5
Budget	6
References	7

Proposals 47

Summarizes basics of proposal, including cost

Executive Summary

Bicycle accidents occur far too frequently in our state, often resulting in permanent injury or even death. Head injuries are the most serious type of injury for cyclists, whether they are killed or just injured. These injuries are expensive both in terms of money and—more importantly—in terms of human suffering.

I propose to study the problem by researching accident statistics and medical journals, interviewing cyclists and cycling organizations, reviewing coroners' records and records of organizations involved in vehicular accidents. I will determine what factors contribute to cyclist head injuries and what can be done to reduce the severity or number of head injuries. I will explore possible solutions to this problem and analyze their merits.

I can accomplish this study and present the results for $600. The benefits of reducing head injuries are fewer injuries and fatalities, less human suffering, reduced medical and insurance costs, and a reduction of the indirect costs in lost time and lost productivity.

Summarizes the problem

An Overview of This Proposal

Thousands of our state's residents are injured or killed each year in bicycle accidents [1, 2]. Those people lucky enough to survive often suffer from serious head injuries. Besides the personal suffering of the injured cyclists and their families, these injuries place a large economic burden on our state's hospital care system.

States writer's qualifications

I have over six years' research experience focused largely on literature and electronic information searches. As a member of a local bicycling organization and a bicycling commuter, I am aware of the issues facing bicyclists.

Summarizes plan for study

I propose to research why bicyclists are injured. I will then analyze how to reduce the number or severity of head injuries.

Justifies need for study by describing problem in detail

Cycling Head Injuries Are a Severe Problem

Each year more than 10,000 state-resident cyclists are injured or killed in traffic accidents [2]. Head injuries are the most commonly reported serious injury, making up approximately 32% of reported serious injuries [3]. Each year nearly 90 residents die in this and neighboring states in bicycle-related accidents (see Figure 1), and 75% of these deaths are due to head injuries [3]. Approximately

20% of those people who survive the accident incur permanent brain damage.

Serious injuries such as head injuries can be expensive to treat. The average one costs over $400,000 in medical costs, related costs, and lost time and productivity [4]. Brain injuries, for example, cost over $1,000 per day to treat, and over $200,000 per year in extended care costs [5]. The lost potential when a young cyclist is seriously brain damaged or killed is incalculable.

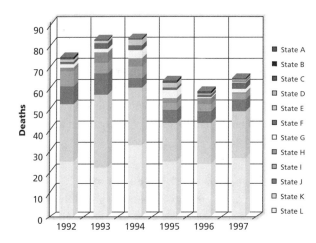

Figure 1: Cycling deaths by year.
Source: Federal Highway Safety Commission, 1997.

Some Real-Life Examples of the Problem

To illustrate the problem of head injuries, here are some case reports from the Hub County coroner's office [6]:

Case #1. A man in his mid-sixties was riding his bike on the shoulder of the East Ridge Highway near Far Hills Avenue (a controlled intersection) heading toward Anytown; he turned on to the slow lane of traffic without signaling, causing the van in that lane to brake abruptly to avoid hitting him. The cyclist then continued over to the fast lane and was struck from behind by another vehicle. Road and weather conditions were excellent. The cyclist died of massive injuries, in particular, head injuries. He was not wearing a helmet.

Case #2. The death resulting from this cyclist/motorist collision involved a 4-year-old girl riding her bike into a "T" intersection when a motorist, observing her approach, nevertheless, thinking he had time, turned in front of her and she ran into the right front of this vehicle. There were no mechanical or environmental factors affecting this accident. The little girl had just recently learned to ride without training wheels and was practicing stops and starts. She died immediately of massive head injuries.

Case #3. A 12-year-old boy was riding his bike on the shoulder of the Lamont Highway against traffic in the late afternoon of a mid-January day. It was already dark. A truck pulled over onto that shoulder of the road to pass a vehicle turning left and struck the cyclist. The bike was not equipped with a light or reflector on the front. The cyclist was not wearing a helmet. The youth suffered multiple traumatic injuries, the most severe being head injuries. Death was instantaneous.

A Problem for Many

Head injuries due to bicycle accidents are a serious problem not only for the injured person. The families of the injured people experience both economic and psychological stress. Medical staff in hospitals must deal with the hundreds of traumatic injuries and deaths each year. The automobile and medical insurance system must bear the burden of the costs of these injuries. Society bears the cost of losing years of productivity from these individuals, who are typically active young people or children with their entire lives in front of them.

Reducing the number or the severity of serious head injuries due to bicycle accidents will certainly benefit all these groups. Clearly, the primary benefit would be fewer needless injuries and deaths in our state. All residents would benefit from reducing the cost of these injuries in terms of the direct medical costs, the insurance costs, and the lost time and productivity. In 2001–2002, the value of lost time and productivity due to serious or catastrophic bicycle accidents was approximately $31 million. On top of that, the direct medical costs and related expenses added another $4.6 million [6].

A study of cyclist head injuries would help find solutions to reducing the number and severity of these injuries. A reduction in bicycle-related head injuries will quickly show up in hospital admission records, which one could monitor on a monthly or semiyearly basis. Furthermore, several organizations including the Federal Bureau of Statistics, our state's Department of Highways, and insurance companies maintain accident statistics, which would reveal a reduction in the number of bicycle-related injuries and deaths.

What the Study Will Do

Details what the study will examine

I will research several aspects of this problem by examining medical, statistical, and Internet-based information combined with interviews with local cyclists and cycling organizations. Specifically, I will address the following questions:

1. What factors affect mortality and injury severity of those involved in bicycle accidents?
2. What are the factors affecting the severity of head injury in bicycle accidents?
3. What changes will reduce the incidence and/or severity of head injuries?

Given the head injury statistics, others proposing to study this problem may jump to the conclusion that mandatory helmet laws are the solution. I do not think one should jump to any conclusions without having studied the full extent of the problem. I plan to examine what factors contribute to bicycle accidents and what factors result in head injury. I will then examine how head injuries can be reduced both by reducing accidents in general and by finding ways to reduce the number and severity of head injuries that result from those accidents.

One may object to this proposal since only 18 people in our state die each year, and only 75% of these deaths are due to head injuries. Indeed these numbers are not large. However, these figures do not take into account the number of head injuries not resulting in death. Nor do these figures really reflect the true cost of head injuries. The average serious head injury costs over $400,000 in medical costs, related costs, and lost time and productivity [6]. On top of these costs, there is a huge emotional cost for the families of those injured or killed.

4

Possible Solutions to Investigate

Defines scope of study

1. Bicycle helmets
   - Laws, compliance with laws
   - Education about helmets
   - Design of helmets
   - Free helmets
2. Restrict bicycles from busy routes or at dangerous times
3. More bike routes and lanes
4. Driver and cyclist education
5. Increased or cheaper public transit

Action Plan

What Has Been Done So Far

Provides detailed plan and progress to date

- I have spoken to a representative of the Federal Department of Highways and obtained national fatality statistics.
- I have spoken with a representative of our state's Office of Statistics and obtained fatality statistics.
- I have done extensive searching of Web-based literature to determine the extent of the problem.

What I Will Do (See Figure 2 for the Time Line I Will Follow)

Describes forthcoming phases of plan

Researching solutions to the problem of cyclist head injuries will take place in four stages:

(1) Literature Review

- I will review paper and Internet-based literature, including medical and statistical literature, to explore methods of reducing the number and severity of head injuries.
- I will examine records in an international cycling accident database on the Internet to determine factors in cycling accidents.
- I will study recent papers on cycling accidents and head injuries including:
  a. the state's Chief Coroner's 1999 Report on Deaths of Cyclists
  b. Dr. Jean Farsight's 2001 report, *The Effectiveness of Bicycle Helmets: A Review.*
- I will study state collision statistics to profile cycling accidents and explore solutions to the problem.

(2) Interviews
- I will consult with local cycling organizations to explore possible solutions to the problem.
- I will interview cyclists to profile local cycling safety concerns.

(3) Expert Consultations
- I will contact auto insurance companies and similar safety organizations to explore aspects of this problem nationally and in neighboring states.
- I will contact the Motor Vehicles Branch in Anystate and similar organizations in other states to explore aspects of this problem.

Provides timetable for each phase of study

Action	Projected Completion Date
Local public library research: statistical information	Oct. 16
Gridley University Library: medical and statistical information	Oct. 21
Internet research: statistical information, research possible solutions	Oct. 25
Cyclist interviews	Oct. 31
State Safety Council and similar associations in other states	Oct. 31
Department of Motor Vehicles and similar associations in other states	Oct. 31
Department of Highways	Oct. 31
Local cycling organizations: Greater Hub City Cycling Coalition and Harbor Wheelers	Oct. 31
Gather and analyze collected information. Analyze possible solutions. Write final report.	Nov. 15

Provides detailed budget

Budget

Internet service	$50
Long-distance telephone calls	$100
Car and gasoline expenses	$100
Requests for information	$100
Printing and computer equipment	$200
Miscellaneous expenses	$50
Total	**$600**

I believe this to be an extraordinarily good value for helping reduce the enormous cost of head injuries due to bicycle accidents each year.

Cites references to provide support for study

References

1. Highway Safety Council: <u>Road Safety Leaflets: Fatality Statistics by Road User Class, 1992–1997</u>. Voorhees, NJ: Highway Safety Council, 1997.
2. <u>Transport Anystate Traffic Collision Statistics</u>, Statistics Anystate [by phone and published data].
3. National Head and Spinal Cord Injury Prevention Program: 1998 Head Injury Statistics. Accessed online at http://www.medi-fax.com/thinkfirst/didja3.html.
4. State Department of Recreation: Serious Head Injuries in Bicyclists, 1999. Accessed online at http://www.gov.on.ca:80/MCZCR/english/sportdiv/communiq/comm09.html.
5. <u>Report of the Anystate Commission on Health Care and Costs, Closer to Home</u>. City, Anystate: State Publishers, 1998.
6. <u>Deaths of Cyclists in Anystate</u>. City, Anystate: Office of the Chief Coroner, 2001.

Proposal 2

LEVERETT LAND & TIMBER COMPANY, INC.

creative land use
quality building materials
architectural construction

January 17, 2006

Mr. Thomas E. Muffin
Clearwater Drive
Amherst, MA 01022

Dear Mr. Muffin:

I have examined the damage to your home caused by the ruptured water pipe and consider the following repairs to be necessary and of immediate concern:

> Exterior:
> Remove plywood soffit panels beneath overhangs
> Replace damaged insulation and plumbing
> Remove all built-up ice within floor framing
> Replace plywood panels and finish as required
>
> Northeast Bedroom—Lower Level:
> Remove and replace all sheetrock, including closet
> Remove and replace all door casings and baseboards
> Remove and repair windowsill extensions and moldings
> Remove and reinstall electric heaters
> Respray ceilings and repaint all surfaces

This appraisal of damage repair does not include repairs and/or replacements of carpets, tile work, or vinyl flooring. Also, this appraisal assumes that the plywood subflooring on the main level has not been severely damaged.

Leverett Land & Timber Company, Inc. proposes to furnish the necessary materials and labor to perform the described damage repairs for the amount of six thousand one hundred and eighty dollars ($6,180).

Sincerely,

G.A. Jackson

Gerald A. Jackson, President
GAJ/ob
Enc. Itemized estimate

Proposal 3

TO: Dr. David McMurray, Director of Marketing
FROM: Robert A. Freund, Contractor—Creative Services Department
DATE: January 6, 2003
RE: Proposal to Develop a Corporate Standards Manual for Darton Corporation

Thank you for asking me to submit a proposal to develop a corporate standards manual. I have had this project in mind for some time now and have been eagerly waiting for the opportunity to work on it. As we discussed in our meeting on December 7, the changes that Darton has experienced over the last few years have strongly impacted its identity. Employees, customers, and the market are all trying to define the new Darton and set it apart from the old Darton. Both the old and the new Darton employees are confused about the usage of Darton brand names, the Darton logo, and other formatting and style issues. It is the perfect time to re-establish a strong identity for Darton, starting with a new standards manual.

The attached proposal outlines the need for and the benefits of a standards manual. It also includes the method I will use to develop the manual, the contents of the manual, costs, schedules, and my qualifications for review by other colleagues if necessary.

Please call me at 770-XXX-1512 if you have any questions. I look forward to hearing from you.

Attachment: Proposal

PROPOSAL
to
Develop a Corporate Standards Manual for Darton Corporation

The following is a proposal to develop a Corporate Standards Manual for Darton Corporation. After reviewing the literature and the outdated standards manual that you gave me at our meeting on December 7, 2002, I developed this proposal to describe to you the process involved in creating your new standards manual. This proposal contains information on the need for a standards manual, the process of developing the standards manual, the contents of the standards manual, the schedule to complete this project, costs to complete this project, and my qualifications to produce a high-quality finished manual. Because I am a former full-time employee of Darton and have now been working for Darton on a contract basis for over a year, I am confident that I can create a standards manual that will accurately reflect the new Darton image and will be a valuable resource for Darton employees.

Need for a Corporate Standards Manual

Since Darton has used a standards manual before, I know that you understand the importance of having a corporate identity that establishes a consistent impression across all media. Because Darton has undergone so many changes in the last few years, it is more important then ever to present a consistent image to the media. There has been a large turnover in personnel since emerging from Chapter 11 bankruptcy in 2000 and the merger with Access Beyond in 2001. Over the past two years, I have seen numerous examples of inconsistent use of Darton identity standards. Some of the incorrect usage is due to the lack of training provided to new employees, and some of it is because so many things have changed yet no new standards have been set. I recently saw a copy of a letter sent from a Darton employee to a customer in which Darton is referred to three times in three different ways: first as Darton Microcomputer, then as Darton Corporation and then the third time as Darton Microcomputer Products, Inc. Not only can inconsistency confuse the customer; it can also leave an impression that Darton doesn't quite have its act together.

Almost everything except the Darton logo has changed. It is time to re-establish the Darton look with new standard formats for logo usage, internal and external correspondence, business cards, forms, signage, press releases, and all media formats. There are new product lines, icons associated with those product lines, and Darton trademarks that need to be used consistently and accurately. Some of the retail cartons, technical manuals, and product literature that I have seen recently do not refer to either the new or the old products consistently. Product names are sometimes written in all caps, sometimes in upper and lower case, and often they are separated from other elements that make up the entire brand name. I have seen usage of the Darton name with an apostrophe, which has always been unacceptable in any situation. As you can see, it is more important than ever that Darton presents a positive and organized impression, especially after its recent well-known financial and organizational problems.

Benefits of a Corporate Standards Manual

How you visually communicate your company to the world, the market, and your clients is an important part of your success. Consistent usage of style and identity gives the world an impression that can be remembered. Once an impression is made to a potential customer through various media materials such as marketing collateral, signage, or the World Wide Web, it should be easily recognized a second time. If the identity elements are not used consistently, the impression may be lost. By creating and using identity standards, we can make Darton more easily recognizable and memorable.

Process of Developing a Corporate Standards Manual

The standards manual that I am proposing to develop will include all identity system elements. I would like to meet with key Darton personnel to establish clear communication objectives. I will then organize and format the manual and design the cover. No new graphics will need to be created, since all the necessary graphics already exist and are archived in the Creative Services Department. New designs will be developed, however, for stationery, presentations, forms and business cards using the current logo with the new Darton name. I propose to also manage the production and distribution of the standards manual and to prepare a presentation for training employees. A permanent Darton employee or myself may be considered to conduct the training session.

I identified three phases in the process of developing the manual: (1) analysis, (2) design, and (3) production. Below are the steps that will be taken in each phase to ensure accuracy and efficiency:

Analysis Phase
1. Research industry visual standards
2. Review competition's standards manuals
3. Attend meetings with key Darton employees to:
4. Define requirements and establish clear communication objectives
5. Identify application items such as stationery, publications, signage, etc.

Design Phase
1. Develop the content and organization of the standards manual
2. Create the format and cover design
3. Finalize basic identity system such as typefaces, colors, etc.
4. Present draft of standards manual for review
5. Incorporate changes as necessary until final approval of standards manual
6. Create presentation material for training session

Production Phase
1. Choose vendor for print production, determine quantity
2. Prepare camera-ready artwork for printer
3. Review proofs and blue lines
4. Supervise prepress, printing and manufacturing of the standards manual
5. Distribute to all employees and set up training session

Description of the Finished Product

This standards manual is for all Darton employees and Darton contractors to use. I propose to use three-ring binders with printed covers and printed tab inserts to separate sections. This will be very useful later when only particular pages or sections need to be updated; rather then reprinting an entire book, only the pages that are changed will need to be reprinted and replaced in the binder. I have estimated that the book will contain between 60–75 pages. Most of the pages will be black and white text, except for those containing graphics such as logos or icons, examples of presentation layouts, or any other standard design element that includes color.

Following is an outline of the sections I plan to include in the standards manual. Some of this may change or sections may be added once I have met with Darton personnel to determine the content.

1. Introduction: will contain brief history of Darton and the proper usage of the Darton name
2. Logo Usage: will contain information on proper usage of the Darton logo and color schemes
3. Product Lines: will contain subsections with information on each product line and icons associated with those product lines
4. Formats: will contain subsections with information on memo formats, fax formats, business stationery, business cards, etc.
5. World Wide Web: will contain information on formatting issues for the World Wide Web
6. Marketing Literature: will contain information on formatting issues for promotional items, print and online documentation, signage, etc.
7. Presentations: will contain standard formats for internal and external presentations

8. Glossary of Trademark names: will contain a list of all trademark names
9. Glossary of Acronyms: will contain a list of Darton and industry acronyms

Project Schedule

The proposed time schedule for this project will be as follows:

January 25	Begin work on project
February 12	Analysis phase complete; begin design work
February 26	Present draft copy of the standards manual to Darton for review
March 8	Incorporate all changes and present 2nd draft for review
March 15	Obtain approval of final copy; design phase complete
March 16	Begin production; deliver artwork to printer
March 22	Proofs from printer reviewed and approved
March 31	Standards manuals delivered to Darton and distributed to all personnel
April 5	Presentations and training sessions begin

My Qualifications

My qualifications for this project are as follows:
- I have been employed at Darton in the Marketing and Communications department as a permanent employee and as a contractor for over five years combined.
- I have extensive experience and knowledge in all design and print production mediums as well as technical and marketing publications development.
- I have successfully completed numerous other contract jobs for Darton.
- I offer competitive pricing.

Costs

I have estimated that it will take approximately 250 hours to complete this project, starting from the day I begin work until I receive the final print copies of the manual. As with the last contract I completed for Darton, my hourly rate is $35 per hour, making the total cost for my services $8,750.

The cost of materials to produce the finished standards manuals can vary considerably depending on the vendor we choose and the quality of materials. I have a number of suggestions and price quotes for the production of the manual, which we can discuss and decide upon during the initial stages of development.

Closing

Thank you for considering me for this project. I hope that you will approve my proposal and consider beginning this project as soon as possible. I am excited about creating this standards manual and as always, I enjoy working with Darton.

Proposal 4

Format
Is the format professional in appearance?
Are headings logical and adequate?
Does the title announce the proposal's subject and purpose?

Content
Is the problem clearly identified?
Is the objective clearly identified?
Does each key element in the proposal support its objective?
Are ideas and claims supported with facts or specific discussion?
Is the proposed plan, service, or product beneficial?
Are the proposed methods practical and realistic?
Are the foreseeable limitations and contingencies identified?
Is the proposal free of overstatement?
Is the proposal ethically acceptable?

Organization
Is there a recognizable introduction, body, and conclusion?
Does the introduction provide sufficient orientation to the problem and the plan?

To: Dr. John Lannon
From: T. Sorrells Dewoody
Date: March 16, 20XX
Subject: *Proposal for Determining the Feasibility of Marketing Dead Western White Pine*

Introduction
Over the past four decades, huge losses of western white pine have occurred in the northern Rockies, primarily attributable to white pine blister rust and the attack of the mountain pine beetle. Estimated annual mortality is 318 million board feet. Because of the low natural resistance of white pine to blister rust, this high mortality rate is expected to continue indefinitely.

If white pine is not harvested while the tree is dying or soon after death, the wood begins to dry and check (warp and crack). The sapwood is discolored by blue stain, a fungus carried by the mountain pine beetle. If the white pine continues to stand after death, heart cracks develop. These factors work together to cause degradation of the lumber and consequent loss in value.

Statement of Problem
White pine mortality reduces the value of white pine stumpage because the commercial lumber market will not accept dead wood. The major implications of this problem are two: first, in the face of rising demand for wood, vast amounts of timber lie unused; second, dead trees are left to accumulate in the woods, where they are rapidly becoming a major fire hazard here in northern Idaho and elsewhere.

Proposed Solution
One possible solution to the problem of white pine mortality and waste is to search for markets other than the conventional lumber market. The last few years have seen a burst of popularity and growing demand for weathered barn boards and wormy pine for interior paneling. Some firms around the country are marketing defective wood as specialty products. (These firms call the wood from which their products come "distressed," a term I will use hereafter to refer to dead and defective white pine.) Distressed white pine quite possibly will find a place in such a market.

Scope
To assess the feasibility of developing a market for distressed white pine, I plan to pursue six areas of inquiry.

Does the body explain *how, where,* and *how much*?
Does the conclusion encourage acceptance of the proposal?
Are there clear transitions between related ideas?

Style and Page Design
Is the level of technicality appropriate?
Is the tone appropriate?
Is the writing style clear, concise, and fluent throughout?
Is the language convincing and precise?
Is the proposal grammatically correct?
Is the page design attractive and accessible?

---

1. What products presently are being produced from dead wood, and what are the approximate costs of production?
2. How large is the demand for distressed-wood products?
3. Can distressed white pine meet this demand as well as other species meet it?
4. Does the market contain room for distressed white pine?
5. What are the costs of retrieving and milling distressed white pine?
6. What prices for the products can the market bear?

Methods
My primary data sources will include consultations with Dr. James Hill, Professor of Wood Utilization, and Dr. Sven Bergman, Forest Economist—both members of the College of Forestry, Wildlife, and Range. I will also inspect decks of dead white pine at several locations and visit a processing mill to evaluate it as a possible base of operations. I will round out my primary research with a letter and telephone survey of processors and wholesalers of distressed material.

Secondary sources will include publications on the uses of dead timber, and a review of a study by Dr. Hill on the uses of dead white pine.

My Qualifications
I have been following Dr. Hill's study on dead white pine for two years. In June of this year I will receive my B.S. in forest management. I am familiar with wood milling processes and have firsthand experience at logging. My association with Drs. Hill and Bergman gives me the opportunity for an in-depth feasibility study.

Conclusion
Clearly, action is needed to reduce the vast accumulations of dead white pine in our forests. The land on which they stand is among the most productive forests in northern Idaho. By addressing the six areas of inquiry mentioned earlier, I can determine the feasibility of directing capital and labor to the production of distressed white pine products. With your approval I will begin research at once.

Proposal Cluster: Request for Proposals

December 5, 2005

**REQUEST FOR PROPOSAL for Services to Create a Master Plan for
Median Nursery School
133 Popham Road
Median, NY 10583**

Median Nursery School seeks proposals from architects for the creation of a master plan for the meetinghouse. Proposals should include a detailed description of the process through which such a master plan would be created, résumés of individuals who would be involved in the work on the master plan, and a fee structure including overall cost for the development of such a plan and expectations for how this cost would be invoiced.

The services requested at this point are limited to the creation of a master plan but might lead to a continuing relationship if and when the Nursery School Board approves going forward with some or all of the plan, and if and when appropriate funds are in place.

Who Are We?

The Median Nursery School is a private, non-affiliated nursery school providing half-day services for approximately forty children from age two through age five, in age-divided groupings. The Building Committee of Median Nursery School has been meeting since last spring to consider issues associated with the physical plant, which was constructed in 1949, has been added onto twice since then, and is used by multiple groups with different functions. As with many institutional buildings, the Nursery School building has grown incrementally over the years. Interior changes (the renovation of classroom spaces, the addition of storage closets, and so on) have been carried out without serious effort to look at the building as a whole. Infrastructure repairs (new roof, changes to the heating system) have been undertaken as necessary, rather than on a regular maintenance schedule.

Who Uses the Building?

The Median Nursery School
- Monday-Friday, 8-12:30: Classrooms, multipurpose room (except Monday), outdoor playground
- Weekly faculty meetings: library
- Parent meetings and other large gatherings, several times each year: Large meeting room

Congregation Beth Shalom (tenant)
- Services every Saturday morning: Large meeting room
- Social hour following services: Multipurpose room
- Various holiday observances throughout the year: Multipurpose room

Community Book Club (tenant)
- Weekly evening meetings: Multipurpose room, one classroom

United Nations Women's Guild (tenant)
- Every other Monday morning: Multipurpose room

What Do We Want to Achieve?

Our committee seeks to commission an architectural plan that will allow us to meet the needs outlined below in incremental steps, as the Nursery School approves the work and is able to raise necessary funding. Although our list of goals is extensive, our intent is that a master plan would allow for the minimum necessary work to address those goals.

- **Kitchen.**
  - Moderately expanded and more efficient work space, with a more open plan
  - Adequate storage for all user groups
  - Hidden trash cans
- **Entrance and downstairs foyer**.
  - A clearly identifiable main entrance
  - A heat-efficient entrance that does not allow cold air in or heated air out
  - Accommodations for hanging coats
  - A more welcoming foyer.
  - More clearly identified entrances to the classroom wing, the multipurpose room, and the large meeting room (upstairs).
- **Access for those with disabilities**.
  - Indoor disability access to the second floor, as well as better toilet access.
- **Additional nursery school needs.**
  - Adequate storage
  - A "sink room," so that children can wash up from outdoor play and clean up special projects like paint without using the kitchen sink.
  - Attractive storage for song books to replace (or repair) the current glass-fronted cabinets, which are mildewed
- **Classroom wing.**
  - Better use of hallway space for storage.
  - More sensible allocation of storage space, so that two corner rooms with multiple windows are not devoted to storage.
  - More thoughtful design of storage space to accommodate multiple user groups and provide each with a sense of ownership.
- **Upstairs foyer and library.**
  - Easily accessible storage for the large items used weekly by Congregation Beth Shalom
  - Display racks for pamphlets and a Nursery School calendar
  - Winter storage for air conditioners and fans.
  - Attractive, efficient storage for both the Nursery School and the Congregation for small items that must be stored upstairs, such as the yarmulkes, prayer shawls, books, and other items now stored by the Congregation in the library.
- **Throughout.**
  - **Drainage**. Effective solutions to our problems with moisture and mildew.
  - **Wiring.** An assessment as to whether our current wiring is sufficient. At minimum, a new circuit for the air conditioners.
  - **Windows.** Are these energy-efficient?
  - **Insulation.** Do we have enough?
  - **Water heating**. Are there alternatives to the single large water heater that keeps water heated all the time?
- **Outside.**
  - Enclosed trash bins

- Shed for building maintenance equipment
- Assessment of condition of parking lot
- Landscape design that would allow the Nursery School to own some space, but would also allow the Meeting to have an uncluttered exterior path into the Meeting House

How Do We Hope to Achieve This?

We anticipate that the process of creating a master plan for Median Nursery School will take approximately four to six months.

We would anticipate the following key dates:

By 1/20/2006:	**Architects submit proposals to committee.**
By 2/15:	Committee selects architect and requests approval from Nursery School Board to engage services of selected architect.
Ongoing, 2/15–5/30:	Architect meets with committee regularly, probably biweekly, to discuss goals, begin to present initial ideas and options, and hear responses.
By 4/30:	Architect meets with Nursery School Board at least twice—once to listen to ideas and concerns, and once to present and discuss initial ideas and possible options and seek responses.
By 5/30:	Architect submits final master plan including options and presents this plan to the Nursery School community or the Board, as decided by the Board.

What Do We Expect the Master Plan to Include?

The most important requirements for the master plan are that it address the goals listed above, that it be capable of prioritized incremental implementation such that each step will take us closer to our goals, and that there be costs attached to each increment (with the understanding that time may increase these costs). Specifically, we would want the master plan to include:

- Conceptual floor plans that provide integrated, coherent solutions for the goals listed above.
- A clear program of incremental, prioritized stages that would allow the Nursery School to undertake the work in steps, addressing our highest concerns first, as the Nursery School approves and as funds are raised, with the understanding that this process may take some years.
- Detailed cost estimates for the work proposed, attached to each incremental step, with the understanding that these estimates would need to be revised as time passes.
- Renderings of any proposed exterior changes.
- An assessment of infrastructure needs, a prioritized list of their urgency, and a maintenance plan for infrastructure going forward.
- A conceptual plan for any outdoor changes proposed.

Sarah L. Rhinesmith
Chair, SNS Building Committee

26 Beechwood Road
Median, NY 105xx

Proposal Cluster: Architect's Proposal 1

mitchell koch architects

145 palisade street • suite 324 • dobbs ferry • new york 10522 • tel. 914.674.0042 • fax. 914.674.2334 • mail@mkastudio.com

January 16, 2005

Median Nursery School
133 Popham Road
Median, NY, 105xx
Re: Master Plan Proposal

Dear Ms. Rhinesmith,

Thank you for the opportunity to submit a proposal (see attached) to provide a Master Plan for modifications and improvements to the Median Nursery School. Working with a group rather than an individual requires certain finesse. It is not only important to address the primary needs of the school; it is equally important to display good stewardship. Together we must make the best use of all that has been provided by and to the school. We are certain the enclosed material will demonstrate that we have the expertise and experience to successfully address the unique goals and requirements of your institution.

Of particular note is the recently completed improvement of a similar nursery school property in Purchase, New York, a project with many similar characteristics. We worked closely with the building committee and Board members on all aspects of the project, including development and maintenance of the budget and allowing the facility to remain in operation as much as possible during the construction. That project involved the delicate knitting together of two additions, housing two new classrooms and a larger multipurpose room, with the historic structures, as well as new building systems, extensive exterior restoration and the enlargement of the kitchen and improvements to existing classrooms. We think that the project enhances the experience of all the members including the children and the staff. In addition, we have prepared several other Master Plans including the Hastings Community Center and the Fieldstondale Mutual Cooperative.

For this project, we have picked a team of consultants chosen for their creativity and responsiveness, and relevant experience. These consultants will participate to develop a Master Plan Report that includes a Scope of Work and design approaches with associated costs (outlined in the proposal). This Team will bring the following complementary attributes to the project:

- Project management and administration expertise
- Experience with projects of very similar scope and complexity
- A reputation for attention to detail
- Principal involvement
- Building code and zoning expertise.

Please call with any questions and thank you for your consideration.
Very truly yours,
Mitchell Koch

Attachment:
Proposal w/ Fee
References
Firm Profile
Images

mitchell koch architects

145 palisade street • suite 324 • dobbs ferry • new york 10522 • tel. 914.674.0042 • fax. 914.674.2334
devensharma@optonline.net • mail@mkastudio.com

PROPOSAL

The intent of this proposal is to outline our services for development of a Master Plan, and give you a sense of how we would work with you to successfully and efficiently accomplish your goals. It should be read in conjunction with the cover letter, firm profiles and project list for a complete picture of who we are and what we can offer the Median Nursery School. Mitchell Koch Architects (MKA) prides itself on its successful client relationships and I urge you to contact the references attached.

METHODOLOGY

MASTER PLANNING PHASE—Thinking Outside the Box

The Median Nursery School has assembled a detailed, but preliminary, list of functional requirements and required improvements. Our first task will be to work with you to develop a comprehensive project Scope of Work that prioritizes both the functional requirements as well as building issues, (architectural and engineering) and establishes associated preliminary estimates of cost. We will then provide a range of viable and creative design alternatives that address the items identified in the Scope of Work.

Master Planning would be organized as follows:

Tasks

- The design team will conduct a site and building survey and conditions analysis to insure that all structural, mechanical, and building envelope issues are identified. A landscape architect will make a site visit so MKA can incorporate their input into our designs. They will also provide landscape designs and presentation drawings. An MEP engineer will review building MEP systems and address immediate and future maintenance issues.
- MKA will review building codes and zoning codes, and ADA requirements.
- Existing Condition Drawings will be field verified and updated.
- We will meet with Nursery School Building Committee both to present findings and to review and confirm functional space requirements
- The Scope of Work, in the form of a matrix, will address architectural, landscape, mechanical and functional items
- The Building Committee will review and comment upon this material.
- We will modify the document to incorporate the comments and resubmit it for approval.
- When BUILDING COMMITTEE approves the Scope of Work and we will present it to the entire Meeting.
- Based upon the approved Scope of Work, we will prepare diagrammatic design alternatives for accomplishing your requirements with associated costs
- We will evaluate the strengths and weakness of the alternatives and make recommendations
- Meet with Building Committee to explain options and recommendations and set priorities and budget
- Building Committee to review and comment upon the material
- Revise material to incorporate Nursery School Board comments
- Develop a schedule for completing the next phases of architectural work based upon the approved scope of work and the Median Nursery School budget and requirements.

- Present to Nursery School Board.
- Incorporate input from Nursery School Board to designs and submit to Building Committee final approval

Deliverables: Scope of Work matrix, alternative design diagrams, (including landscaping), with a preliminary estimate of associated costs and recommendations.

TEAMWORK

We believe that the key to success of any project is the quality of the participants and the ability to work together with you, the stakeholders, as a team. We have selected the sub-consultant team based upon their creativity and responsiveness and our previous experience working together. The following is the proposed core team. Please note that other consultants can be added upon your direction as required, for additional fees. For example, you may desire a lighting or acoustical engineer, or between us we may determine that thermal imaging will aid in the energy conservation evaluations.

The Core Team

Architecture Mitchell Koch Architects, Dobbs Ferry, NY
Landscape Architecture Sanok Design Group, Brewster, NY
MEP Engineer Sigma Psi Consulting, Malta, NY

SCHEDULE

We will work with the schedule in your RFP to complete the work and are available to begin per your requirements.

FEE

MKA will bill for its work on an hourly basis (see rates below) with a maximum amount capped at fifteen thousand dollars ($15,000). A retainer of three thousand dollars ($3,000) will be required to begin. In addition to our architectural and master planning work the this amount includes:

- A site visit, report and preliminary cost estimate by the mechanical and electrical engineer
- A landscape architect will address site issues
- Materials for presentation to the Friends (However, fundraising materials can be provided at an additional cost)
- A total of six meetings with the Committee and Meeting as a whole. (Based upon our experience 6 meetings will be sufficient for this phase, however, we are happy to attend additional meetings as required, to be billed at our hourly rates.)

Rates

Principal	$150
Architect	$90
Interior Designer	$50
Junior Architect	$50
Administration	$50

Reimbursible Expenses 1.1 These are outside the cap

Note that should you select Mitchell Koch Architects, your signature on this proposal with a retainer can serve as an Agreement for us to begin work.

_____

Representative, Median Nursery School

mitchell koch
architects

145 palisade street • suite 324 • dobbs ferry • new york 10522 • tel. 914.674.0042 • fax. 914.674.2334 • mail@mkastudio.com

FIRM PROFILES
Mitchell Koch Architects, Dobbs Ferry, NY
www.mkastudio.com
The firm works closely with clients, responding to the social and architectural context of project site. Committed to discovering creative design solutions and to bringing them to fruition on time and within budget.

Mitchell Koch, AIA, Principal in Charge, Designer
Mitchell Koch will be directly responsible for this project. Mitch was a contractor and master carpenter for fifteen years prior to getting his Architectural Degree from Pratt Institute in 1991. Since opening his own firm in 1998, his approach remains 'hands on', working directly with the client and contractor on each project. Mitchell Koch Architects' work is comprised of a variety of institutional, commercial and residential projects. Most recently, MKA has completed the restoration of the historic Friend's Meeting House in Harrison, N.Y. and Master Plans for the Community Center for the village of Hastings on Hudson and the Fieldstondale Mutual Housing Cooperative.

Prior to founding the firm, Mitchell was project architect for the design and construction of new academic buildings at the Pembroke Hill School in Kansas City and the extensive renovation of the Hastings on Hudson Village Hall. With Kapell and Kostow Architects in New York City he was project architect for numerous institutional, residential and commercial projects including NYU Dental School, the facilities of HBO Studio Productions and Planet Hollywood.

In addition to his practice, Mitchell Koch has been a professor at Fordham/Marymount and serves on the Architectural Review Board in Hastings-on-Hudson a plays the violin in a quartet.

Roger Sommerfield, Project Architect
Roger Sommerfield received his Architectural Degree from Pennsylvania State University. In the role as project architect, he works closely with the principal, Mitchell Koch, as well as the Client to provide planning and design solutions that address the client's goals. He recently completed the design for the Liberty Bus Lines operations center, the restoration of an elegant turn-of-the-century home in the Rivertowns, and the completion of the punch list for the Purchase Friends Meeting. Roger is experienced in developing and monitoring budgets and coordinating the work of engineers and contractors.

Prior to his work at MKA, Roger was with a large firm where he worked on a variety of structures ranging from high-rise residential buildings to large institutional facilities. Roger has been involved with several sustainable buildings, and brings his "green" architectural knowledge to the firm.

Kathleen Boyd, Interior Designer
Kathleen Boyd received her degree in Interior Design from Marymount College 1991 and has worked in architectural offices since then. For the Median Friends Meeting she will be involved in the administration, coordination and preparation of the Master Plan Report

Sanok Design Group, Brewster, N.Y. /Landscape Architect
James Sanok, Principal

James Sanok, founder of Sanok Design Group, SDG established the firm in 1998 to provide professional services in landscape architecture for institutional, residential and commercial properties. The firm's work ranges from small landscape design projects to large-scale site development and land planning projects. SDG strives to provide integrated design services working in close collaboration with architects and engineers, seeking solutions that are compatible both with their natural and man-made surroundings. In addition, James is committed to working within the constraints imposed by the budget and schedule. James is the principal-in-charge of master planning, design development, construction drawing and specifications, and construction administration for the firm. He develops the initial design concepts for all projects and maintains an oversight on each, from initial site analysis through program, design development and construction documents, to assure that the intent of each design is realized.

Prior to starting his own firm, Mr. Sanok was with Peter Gisolfi Architects where he was responsible for overseeing the landscape architecture department. His projects included the Horace Mann School, the Masters School in Dobbs Ferry and the Castle at Tarrytown, as well as the parking design for the new Dobbs Ferry library.

Recently Sanok Design Group and Mitchell Koch Architects completed the planning phase for the alteration and renovation of building entrances and landscape of a cooperative housing complex in the Bronx. This project addressed recreational uses, site circulation and safety issues, public space and enhancing the sense of the community- critical issues for the SFM.

On this project Mr. Sanok will work closely with MKA on site analysis; in particular the development of welcoming public spaces, drainage issues, safe vehicular circulation and as required, the relationship of the buildings with their surroundings.

Mr. Sanok received a Bachelor in Landscape Architecture from the College of Environmental Science and Forestry at Syracuse University in 1994.

Sigma Psi Consulting, Malta, N.Y./ Mechanical, Electrical, Plumbing Engineers,
Paul Martin, Principal

Mr. Martin founded Sigma Psi in 1997 to offer professional services to architects, governmental agencies and private and public building owners. Prior to that he had over twelve years of experience with other A/E and M/E offices.

The name Psigma Psi was chosen as a reminder of how dependant the practice of engineering is on mathematics and physics. Psigma Psi's approach involves the evaluation of many ideas, a determination of constructability and a review of costs. All variables are then applied to quantified, common sense design solution recommendations

The firm's experience includes historical buildings, schools and colleges, hospitals, and computer date centers. The professional staff members have specialized certifications in the following areas: Plumbing Engineering, Indoor Air Quality, Energy Management, Fire Protection and Lighting Design.

Proposal Cluster: Architect's Proposal 2

January 19, 2005

Sarah Rhinesmith
Chair, MNS Renovations Committee
133 Popham Road
Median, NY 105xx

Professional Services Proposal for Median Nursery School Master Plan

Dear Ms. Rhinesmith,

As per your request for a proposal, following is an outline of professional services to be provided for the Median Nursery School. The intent of the master plan effort will be to investigate the needs, remedies, and estimated costs associated with a variety of planning and deferred maintenance concerns outlined in your letter of December 5, 2005. The utmost effort will be made to be attuned to the spirit and policies of the Nursery School in creating this plan.

This proposal is divided into three parts:
- A general summary of items to be considered.
- Master planning stages.
- Professional team information, limits on scope of services and fees.

Items to be considered:
- Programmatic needs of the different groups using the Nursery School building.
- Kitchen renovation and possible relocation.
- Improved entrance, downstairs foyer, and circulation from entrance to other areas of the building.
- Access for those with disabilities to the second floor and other areas of the building.
- A well functioning multipurpose room.
- Classroom wing analysis.
- Means of furthering the functional aspects of the upstairs foyer & library.
- Improvements to and maintenance of the mechanical, plumbing and electrical systems.
- An analysis of the building envelope and energy efficiency.
- Site & parking lot analysis.

Master Planning Stages:
- We will prepare background drawings as necessary to generally document existing conditions relevant to the proposed scope of work.
- We will gather information and clarify programming and space planning needs.

Allen D. Ross Architect 29 Mill Street, PO Box 182, Clintondale, NY 10530. 845.883.5959 Fax 845.883.9572

- We will provide a general analysis of HVAC, plumbing and electrical systems. Our engineers will rely largely on visual observation and interviews with staff, Faculty, and administrators of the Nursery School familiar with these areas.
- We will complete a preliminary review of code matters as they relate to the proposed alterations, including building code, life safety code, and the Americans with Disabilities Act.
- We will examine zoning issues including the possible need for a setback variance.
- We will work with our engineers and cost estimator to summarize deferred maintenance concerns and suggest priorities and remedies.
- We will investigate options and prepare concept drawings, both floor plans and elevations, to illustrate varying solutions to the Nursery School's concerns. Upon selection of design options an outline of prioritized stages of work will be prepared.
- We will prepare preliminary budget estimates for each item of work identified in the process. These estimates will either be on a square foot or other unit basis or be from our cost estimator or other contractor.
- We will review options, findings, and preliminary conclusions every two weeks with the Building Committee with two additional meetings with the larger Nursery School community. The process will conclude with the preparation of a report summarizing conclusions and a presentation to members of the Nursery School community as the Building Committee deems appropriate.
- The schedule will follow that outlined in the RFP with a final presentation of the master plan at the end of May, 2006.

Professional team information, limits on scope of services, and fees:

Team members:

Architectural: Krist Dodaro, R.A.
Structural: David E. Seymour, P.E.
HVAC/Plumbing/Electrical/Fire Protection: Tucker Associates Consulting Engineers, LLC
Surveyor (as required): Ward Carpenter Engineers, Inc.
Civil (as required): Petruccelli Engineers
Cost Estimating: Murphy Brothers Contracting, Inc.

All work to be coordinated by Allen D. Ross, AIA.

Limits on scope of services:

In order for the work outlined above to take place measured drawings of the existing building and a survey of the property are necessary. These will be required in order to determine code and zoning compliance. If these are not existing then a survey would be obtained with the surveyor to be contracted directly by the Nursery School; measured drawings of the existing building would be provided by us as an additional service at the hourly rates shown below.

Allen D. Ross Architect 29 Mill Street, PO Box 182, Clintondale, NY 10530. 845.883.5959 Fax 845.883.9572

The existing floor framing of the Nursery School building is reinforced concrete. If exploratory holes are required in order to determine reinforcing and strength of beams and slabs, this cost would be billed as an additional service.

Because of limitations imposed by our insurance carriers, we cannot assume responsibility for the investigation or analysis of hazardous materials anywhere on the premises. This would be done by a separate consulting office to be contracted directly by the Nursery School. I would be able to recommend a source for this work.

Fees:

Our fee to complete this work will be charged on an hourly basis, with a not-to-exceed figure of $10,000 plus reimbursables. Only those hours worked will be billed. Reimbursable expenses will be billed at 1.1 times actual costs, and will include the customary items such as mailing and shipping expenses, and printing and reproduction costs. Invoices will be issued monthly as the work progresses. Hourly billing rates for architectural services shall be as follows:

- Principal $115.00 per hour
- Senior Designer $85.00 per hour
- Draftsperson $65.00 per hour
- Clerical $40.00 per hour

The hourly billing rates for engineering and estimating services are attached to the associated résumés.

Our project team is enthusiastic about working with you on the master plan and we welcome the opportunity to address the multifaceted requirements of the Nursery School. I look forward to meeting with the Building Committee to further describe our qualifications and work history as well our thoughts about this particular project. Please also note that many of the terms in this proposal are flexible and can be modified as necessary by the needs and wishes of the Building Committee.

Sincerely,

Allen D. Ross, AIA

Allen D. Ross Architect 29 Mill Street, PO Box 182, Clintondale, NY 10530. 845.883.5959 Fax 845.883.9572

Proposal Cluster: Contractor's Proposal

September 28, 2005

Ms. Sarah Rhinesmith
Median Nursery School
Median, NY 105xx

PRELIMINARY PROPOSAL

(1) Excavate and build a two-story addition on the back (west) side of the building, building out from the Library and Meeting Room Foyer on the second floor, with an elevator. Align the addition with the current back wall of the Meeting Room. In this addition, create storage on the first floor for the Nursery School on the ground floor, and expanded Library and Storage on the second floor. $ 210,000.00 ± 10%

Fallback: Build out only second floor to create storage, without excavation. $ 140,000.00 ± 10%

(2) Turn the stairs around to make the entry to the upstairs Meeting Room more apparent. Use the same path that the current stairs occupy. Make the entry to the bathrooms at the back of the corridor. $ 20,000.00 ± 10%

(3) Move the Kitchen back, gut and reconstruct. Make a passageway along the front window wall from the Entryway to the Multipurpose Room (where the Kitchen sink is now). Move the Kitchen back and close off the current passageway to the Multipurpose Room. Rebuild the Kitchen with more efficient storage and new appliances. Supply of appliances, and supply or installation of Kitchen fire suppression system not included. $ 90,000.00 ± 10%

Fallback: Gut the current Kitchen and rebuild with efficient storage and new appliances. Supply of appliances, and supply or installation of Kitchen fire suppression system not included. $ 70,000.00 ± 10%

The following items are not included in the Proposal:
- Temporary fire and heat protection, which may be required by your insurance company.
- Blasting, rock chipping, rock removal, and drilling or pinning for foundation, if required is not included.
- Tile is to be installed on a straight joint. Diagonal, pattern, inlay, or border is not included.

Murphy Brothers Contracting, Inc., 416 Waverly Avenue, Mamaroneck, NY 10543
Office (914) 777-5777 / (203) 629-1291 Fax (914) 777-6658 www.murphybrothers.com

MURPHY BROTHERS CONTRACTING

Painted and Finished Characteristics:
Murphy Brothers Contracting, Inc.'s Hallmark is satisfied customers. Therefore, it is for this reason that we accentuate the need for our customers to completely understand the characteristics of painted, stained, or natural finishes. A situation may exist whereby the cabinetry, trim or flooring in your home will dry out or pick up moisture. Rough framing will also contract and pull adjoining members (sheetrock, exterior soffits, etc.) with them. In either event, the expansion or contraction of the joints can cause the paint finish to fracture at the joints. This condition is not in any way considered defective workmanship or materials, nor will it affect the stability of your woodwork or finish in general. If patching is necessary, it will be done on a Time and Material basis. We cannot be held responsible for natural sap, tannin oil, etc. excretions from any wood species.

Asbestos Clause:
Murphy Brothers Contracting, Inc. and all subcontractor's scope of work shall not include the identification, detection, abatement, encapsulation or removal of any toxic, hazardous or radioactive waste substance, material, chemical, compound or contaminated material including asbestos and polychlorinated biphenyl (PCB), lead or any other hazardous substances. In the event that Murphy Brothers Contracting, Inc. encounters any such products or material in the course of performing our work, we shall have the right to discontinue our work and remove all our employees and subcontractors from the project until no such products or material, nor any hazard exists, as the case may require, and Murphy Brothers Contracting, Inc. shall receive an extension of time to complete our work hereunder and compensation for delays encountered as a result of such situation and correction. It will be the homeowners' responsibility to test for asbestos, lead, or similar hazardous substances before, during and after construction

The Owner acknowledges that Contractor does not hold any special license, permit, authorization or approval, and is not otherwise recognized by Laws and Regulations as a person or entity permitted to handle, generate, transport, treat, store or dispose of any hazardous material.

Concealed Conditions:
This contract is based solely on observations the contractor was able to make with the structure in its current condition at the time the work was bid. If concealed conditions are discovered once work has commenced which were not visible at the time this proposal was made, the contractor will stop work and point out these unforeseen concealed conditions to the owner/architect so that the owner/architect, and contractor can review and decide if they need to execute a Change Order for any deductive or additional work. Not applicable to Time and Material projects.

Murphy Brothers Contracting, Inc., 416 Waverly Avenue, Mamaroneck, NY 10543
Office (914) 777-5777 / (203) 629-1291 Fax (914) 777-6658 www.murphybrothers.com

Deviation from Scope of Work in Contract Documents:
Any alteration or deviation from the scope of work referred to in the contract documents involving extra costs of materials or labor will be executed upon written change order issued by the contractor and signed by the contractor and owner prior to the commencement of additional work. This Change Order will become an extra charge over and above the lump sum contract amount referred to at the beginning of this contract. Not applicable to Time and Material projects.

Supplied by Owner:
If Owner is to furnish any materials or equipment for installation by the Contractor, Owner represents that the materials are either presently on hand at the locations specified or will be made available by the Owner for the Contractor at agreed locations sufficiently in advance of when they are required for installation so as to cause no delay in performing the work. Murphy Brothers Contracting, Inc. shall not be held responsible for specifications, ordering, delivery, time delays due to material delay, defects, or replacement of any items supplied by the Owner.

Final Payment:
Balance of contract amount is due upon Substantial Completion of all work under contract, "Substantial Completion" is defined as the point at which the building/work is suitable for its intended use, or the issuance of an occupancy consent or final permit sign-off from the Building Department, whichever one of the aforementioned events occurs first. Owner may hold back 200% of the value of all punch list work from final payment to contractor to assure that all punch list work is performed in a timely manner. There will be a $75.00 fee for any returned checks.

Payment of Change Orders: 100% of payment for each Change Order is due upon completion of change order work and submittal of invoice by contractor for this work. Not applicable to Time and Material projects.

Murphy Brothers Contracting, Inc., 416 Waverly Avenue, Mamaroneck, NY 10543
Office (914) 777-5777 / (203) 629-1291 Fax (914) 777-6658 www.murphybrothers.com

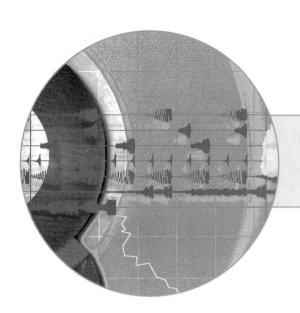

Abstracts

Abstracts are brief statements that precede articles or reports in scholarly or scientific journals. Abstracts are often simple summaries of the most important results of the research report that follows, or the most important points of the article that follows. Many abstracts follow a more structured format and include a claim that the work that is reported in the article is important, a selective summary of previous research on the topic, an argument that the previous research is incomplete for some reason, and a statement of purpose and brief description of the research being presented. For examples of abstracts, see Report #3 (Research Report) or Case Study #1.

Abstracts that precede reports or proposals in business are more precisely called **executive summaries** and serve a slightly different purpose: they are designed to provide a snapshot of the document to follow for busy executives, who may not have time to read the whole report. Executive summaries generally summarize the background and the findings or recommendations in the document that follows and, for a proposal, may include the qualifications of the writer, while only occasionally pointing to the methodology or data sections. For an example of an executive summary, see Proposal #1.

Quick Reminders for Writing Abstracts

- **Include only the necessary information.** The point of an abstract is to make a case to the reader for why the article that follows is important and should be read; the point of an executive summary is to provide an overview of the proposal, findings or recommendations. Keep your purpose in mind.
- **Keep it brief.** Abstracts should typically be no more than 5% to 10% of the length of the document that follows.

Abstract 1: Report

Summarizes content of report

Includes keywords for ease of electronic retrieval

States what is omitted from report

ABSTRACT:
This report summarizes the technical effort of the Active Cooling for Enhanced Performance (ACEP) program sponsored by NASA (NAS3-27395). It covers the design, fabrication, and integrated systems testing of a jet engine auxiliary cooling system, or turbocooler, that significantly extends the use of conventional jet fuel as a heat sink. The turbocooler is designed to provide subcooled cooling air to the engine exhaust nozzle system or engine hot section. The turbocooler consists of three primary components: (1) a high-temperature air cycle machine driven by engine compressor discharge air, (2) a fuel/air heat exchanger that transfers energy from the hot air to the fuel and uses a coating to mitigate fuel deposits, and (3) a high-temperature fuel injection system. The details of the turbocooler component designs and results of the integrated systems testing are documented. Industry Version-Data and information deemed subject to Limited Rights restrictions are omitted from this document.

Abstract 2: Proposal

Abstract

South Florida has been a popular immigration destination for people of African descent since the turn of the century. In today's Broward County, a Black person could be Barbadian, or West Indian, or Belizean, or Haitian, or Cuban, or…all of the above! 25% of Blacks are foreign-born, and students may speak any of 79 languages. As a result, its African-American community is a unique patchwork of histories, traditions, and cultures. In response to the compelling need to collect, preserve, study and interpret such heritage and experience, the Broward County Library (BCL) is building the African-American Research Library and Cultural Center (AARLCC). It will be the first public library to portray the complexity of a community made so diverse by the young, large, strongly-rooted immigrant population from the Caribbean, South America and African countries. This 52,000 sq.ft. research center will open in 1999 as a model institutional hub, harnessing the efforts of several colleges/universities, scores of cultural organizations, and the local community "to enhance the understanding of the African-American experience through study, research, and human interactions by the genera/public, students and scholars…" The new home of an existing Black Heritage Collection, AARLCC will house 75,000 books, documents, AV and electronic materials focusing on Caribbean and West African studies. A Community History Project is developing a local collection of significant African Americana featuring the local pioneers' papers, oral histories and artifacts documenting Black history in South Florida. The 300 seat auditorium and museum-level exhibit space will provide a forum to exchange ideas and cultural value in one of the nation's most multiethnic region. Half of AARLCC $10 million cost is funded by the Broward County government, and the Broward County Library has launched a $5 million capital campaign to fund the other half. As a campaign catalyst BCL seeks a $600,000 NEH Challenge Grant to be matched by $1.8 million: $1.5 million in direct support to meet immediate needs for construction, $300,000 for acquisition of humanities materials, and $600,000 to establish an endowment supporting programming in perpetuity. Expected long term impact of this grant includes: the creation by Florida Atlantic University of an African-American & Caribbean Studies Program based at the Center, an electronic link between AARLCC and the special collections of Florida's four Historically Black Colleges, and new programming activities such as the State of the Race Annual Conference, masters residencies and workshops for children and adults, scholars/authors series, and interpretive programs displaying the community's culture through the expression folklore, dance, theater, film and literature. In a very young county where the challenges of multiculturalism, demographic explosion, and boom/blight economics have many implications for humanities organizations, AARLCC will be the cornerstone of BCL's 1998–2003 strategic plan. It will give the focus, space, new technologies, and leverage necessary to address key issues through the power of humanities initiatives. Standing at the heart of the historically Black neighborhood, next to the business and cultural district, this "bridge between cultures" will foster the mutual knowledge/understanding critical to heed the national call toward healing and reconciliation between races. Its tradition of innovative partnerships and cultural programming leadership give BCL the expertise this project demands. Enjoying strong financial/community support and the recognition that came with winning the 1996–97 Library of the Year Award, BCL is well-placed to raise the $5 million capital goal. The campaign is being launched on three fronts: the Corporate/Private Fundraising Committee, supported by the Broward Public Library Foundation seeks to raise $3 million from businesses, private foundation and local philanthropy; the Citizen's Committee is raising $1 million through the efforts of grass-root organizations; and another million is being sought by way of government grants and municipal challenges. Although $3 million have already been raised, the NEH Grant is critical to attract dollars from national philanthropy in a region of scarce private and corporate foundations and to spur strong financial backing from potential partners in the activities of AARLCC

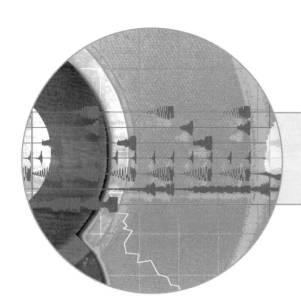

Instructions and Procedures

Instructions and procedures are strictly functional documents. Their purpose is to help somebody do something—and that means that the audience would probably prefer to be doing something rather than reading something. So instructions and procedures need to be user-centered, task-oriented, and user-tested documents. They have to be as brief as utility will permit, and, because people will skim them rather than read them, they need to be well-organized and visually-oriented. They need to be clear and truthful, since they may play a part in warranty and legal issues later on.

Quick Reminders for Writing Instructions and Procedures

- **Familiarize yourself with the product or task.** Before you tell someone how to do something, you have to know how it's done.
- **Don't assume expert users.** Try to anticipate where people will get confused.
- **Test the documents.** Find some real users and see if they can use the documentation, adjust accordingly.
- **Use pictures, diagrams, and tables.** People don't like to read instructions. So use as little text as possible, and rely on visual cues as much as possible.

Instructions 1

Annotations (left margin):
- Attracts user's immediate attention
- Table provides cues for decision-making
- White space provides breathing room
- Headings announce each procedure
- Each step is listed and numbered
- Warning is highly visible (to meet legal requirements)

Installing the Md4 Imaging Board   3

Installing the Md4 Imaging Board

Caution   If you do not properly install the Md4 board, you may be unable to boot your computer. Read and follow the procedures in this guide carefully.

Table 2: Md4 Installation

If...	Then...
You are installing the Md4 board in a 715 workstation	Follow the installation procedure below
You are installing the Md4 board in another workstation	See the installation documentation

To unpack the Md4 board

1. Remove the Md4 board from its package.
2. Remove the protective plastic sleeve slowly to protect the board against static discharge.

To install the Md4 board

1. Turn off the power to your computer.

   Leave the computer plugged into a grounded power outlet so that the power cord serves as a ground for the computer.

WARNING   If you leave the computer turned on, you could get an electrical shock and/or cause damage to your computer's components and the Md4 board.

Instructions 2

Water and Ice Dispensers
(on some models)

Cut Hazard
Use a sturdy glass when dispensing ice or water. Failure to do so can result in cuts.

Depending on your model, you may have one or more of the following options: the ability to select either crushed or cubed ice, a special light that turns on when you use the dispenser, or a lock option to avoid accidental dispensing.

The ice dispenser
Ice dispenses from the ice maker storage bin in the freezer. When the dispenser lever is pressed:

- A trapdoor opens in a chute between the dispenser and the ice bin.
- Ice moves from the bin and falls through the chute.
- When you release the dispenser lever, the trapdoor closes and the ice dispensing stops. The dispensing system will not operate when the freezer door is open.

Some models dispense both cubed and crushed ice. Before dispensing ice, select which type of ice you prefer. The button controls are designed for easy use and cleaning.

- For cubed ice, press the CUBE button until the red indicator appears in the window above the CUBE button.

- For crushed ice, press the CRUSH button until the red indicator appears in the window above the CRUSH button.

For crushed ice, cubes are crushed before being dispensed. This may cause a slight delay when dispensing crushed ice. Noise from the ice crusher is normal, and pieces of ice may vary in size. When changing from CRUSH to CUBE, a few ounces of crushed ice will be dispensed along with the first cubes.

To dispense ice:
1. Press button for the desired type of ice.
2. Press a **sturdy** glass against the ice dispenser lever. Hold the glass close to the dispenser opening so ice does not fall outside of the glass.

IMPORTANT: You do not need to apply a lot of pressure to the lever in order to activate the ice dispenser. Pressing hard will not make the ice dispense faster or in greater quantities.

3. Remove the glass to stop dispensing.

NOTE: The first few batches of ice may have an off-flavor from new plumbing and parts. Throw the ice away. Also, take large amounts of ice from the ice bin, not through the dispenser.

The water dispenser
Chilled water comes from a container behind the meat drawer. It holds approximately $1\frac{1}{2}$ quarts (1.5L).

When the refrigerator is first hooked up, press the water dispenser bar with a glass or jar until you draw and discard 2 or 3 quarts (1.9 to 2.8 L). It will take three to four minutes for the water to begin dispensing. The water you draw and discard rinses the tank and pipes.

Allow several hours to chill a new tankful.

IMPORTANT: The small removable tray at the bottom of the dispenser is designed to catch small spills and allow for easy cleaning. There is no drain in the tray. The tray can be removed from the dispenser and carried to the sink to be emptied or cleaned.

To dispense water:
1. Press a glass against the water dispenser lever.
2. Remove the glass to stop dispensing.

NOTE: Dispense enough water every week to maintain a fresh supply.

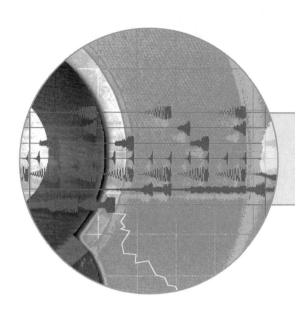

Descriptions and Definitions

Definitions are keys to understanding. If the same word means different things in the same context to different people, communication is impossible. As a writer you have to ensure that the words you use have distinct meanings for your audience. You do this by using common words in their prevailing sense and technical words precisely. If there is a chance your readers will not know the meaning of a word, define it. Don't vary terminology for the sake of variety.

A description puts an image into words, although descriptions are often accompanied by photographs or diagrams to help the reader visualize the object. Your goal is to help the reader see the object you are describing completely enough to understand its parts and how they relate to form the whole thing.

Quick Reminders for Writing Descriptions and Definitions

- **Select details carefully.** Select the details of the object that matter to the audience.
- **Select language carefully.** Use terminology precisely and choose adjectives carefully. Eliminate any language that is not necessary to understanding the object.

Description 1

Provides functional definition

Rivets are used to permanently join sheet-metal or sheet-plastic parts together. They are frequently used when some degree of flexibility is desired in a joint, as when joining the ends of a belt to give a continuous loop. The most common rivet shapes are shown in the figure below. Rivets are made of soft copper, aluminum, or steel. To join two pieces, a hole, slightly larger than the body of the rivet, is drilled or punched in each piece. A rivet is inserted through the holes, and a head is formed on the plain end of the rivet using a hammer or, preferably, a riveting machine. The hammering action swells the body of the rivet to fill the hole.

Notes specific type that might be commonly used

"Pop" rivets, [also] illustrated below, are useful in the lab. These can be installed without access to the back side of the joint.

Source: Building Scientific Apparatus by John H. Moore and Christopher C. Davis. Copyright © 2002. Reprinted by permission of Perseus Books Publishers, a member of Perseus Books, L.L.C.

Visuals support understanding

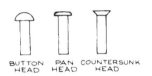

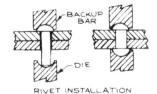

RIVET INSTALLATION

"POP" RIVET INSTALLATION

Description 2

The *arm support assembly*, which is attached to the drive shaft, is the main turning mechanism of the filament-winding machine. This support assembly consists of three main components: the arm support itself, the bearings, and the spring plungers. The bearings and the spring plungers serve as a connection to the plate arm.

Arm Support Assembly

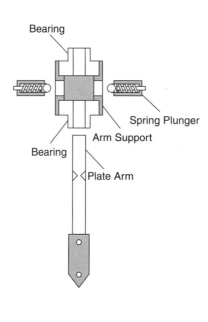

Description 3

Description of Test Equipment

The combination of the large, custom-made calorimeter and the high-power battery cycler is a one-of-a-kind piece of special purpose test equipment for battery calorimetry. Here, we provide a description and the capabilities of each component.

Calorimeter Description

The calorimeter is of the heat conduction type and is based, in part, on a commercially available isothermal calorimeter (CSC Model 4400 Isothermal Microcalorimeter). Heat-conduction calorimeters sense heat flux between the sample and a heat sink. The heat sink is the enclosure containing the sample and is fabricated with aluminum surrounded by an isothermal bath. If the sample is hotter or colder than the heat sink, heat flows between the heat sink and the sample. In actual practice, the thermal conductivity of the path between the sample and the heat sink is matched to the expected heat flow so that the temperature difference between the sample (the battery, in this case) and the heat sink is minimized. The temperature of the heat sink is kept constant and the entire calorimeter shielded from its surroundings by a constant-temperature bath. The temperature control of the heat sink, together with proper matching of the thermal conductivity of the path between the sample (or measurement cavity) and the heat sink, renders a passive isothermal measurement condition.

The large, custom-made battery calorimeter is pictured in Figure 2 and a block diagram of the calorimeter with a listing of critical components is shown in Figure 3. The measuring unit of the calorimeter includes a 39 cm long, 21 cm wide, and 20 cm high aluminum enclosure connected to a large aluminum heat sink via heat

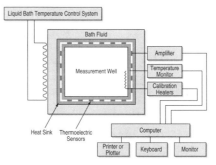

Figure 3. Block Diagram of the Heat Conduction Battery Calorimeter

Figure 2. Large Cavity Battery Calorimeter

flow sensors (semiconductor thermoelectric devices) that are located between the heat sink and the sample cavity. The bath temperature operating between $-30°C$ to $+60°C$ is controlled with a stability of $0.001°C$. For calibration purposes, the measuring unit also incorporates electrical heaters, allowing for heat input at rates of 1 to 80 W. The measuring unit is designed so large-gauge leads, which must be connected to the sample battery for charging and discharging experiments, achieve thermal equilibrium. The large-gauge leads generate a negligible amount of heat even at very large currents. Further, they are attached to the isothermal aluminum enclosure, resulting in an insignificant impact on the accuracy of heat generation data obtained for battery modules. The battery temperature (internal or surface) in the calorimeter is measured with an accurate platinum RTD.

The measurement cavity can be either dry or filled with a dielectric, inert heat-transfer fluid. The air in the dry chamber can be stirred with small fans and the liquid-filled chamber is stirred with constant-speed stirrers to speed heat transfer from samples to the measurement chamber walls. The small amount of heat added for mixing is taken into account in calculations. The time response of the calorimeter is not affected by stirring within the measurement cavity. However, the time required to reach a steady-state response for a battery at constant power will be shortened significantly by stirring due to improved heat transfer between the sample and the calorimeter. Even under the best circumstances, the time constant for a typical large battery module will be much longer than the response time for the calorimeter.

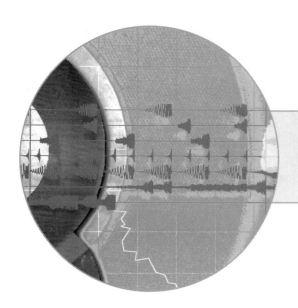

Reports

A report is any document that offers specific information in response to a situation, request, or problem. Reports, which typically provide either description or analysis, are one of the most common forms of business writing. Almost all employees are asked at some point to write one or more of the following kinds of reports:

- periodic or progress reports, that keep other employees or managers up to date on an employee's activities or day-to-day business;
- investigative or exploratory reports that examines a particular circumstance and sometimes proposes solutions to a problem;
- policy reports that examine procedures and sometimes recommend new ones;
- feasibility reports that look at the benefits and risks of a proposed action;
- evaluative reports about themselves or the people they manage.

The typical pattern for a report is problem/solution or context/recommendation—although sometimes, as in the case of meeting minutes or travel reports, a report is simply a description of what happened. Reports vary in length from a single page (or computer screen) to tens or even hundreds of pages. Longer reports are often divided into sections written for different audiences: an executive summary for the decision makers; an analysis for engineers; a product descriptions for marketers, and so on.

Quick Reminders for Writing Reports

- **Understand the purpose of your report.** Are you writing a descriptive report? An analysis? Is your report one of a series issued periodically, or is it a one-time document? Are you examining policy, progress, or feasibility?

- **Understand what your audience wants.** Your reader may want simple information or an analysis of a complex situation. They may or may not want specific recommendations.

- **Make the structure clear.** State at the beginning how the report is organized. Provide recommendations in a bulleted or numbered list rather than embedded in a narrative.

Report 1: Progress Report

<table>
<tr><td></td><td colspan="2" style="text-align:center">MEMORANDUM</td></tr>
<tr><td></td><td>Date</td><td>24 February 2004</td></tr>
<tr><td>**Uses memo format**</td><td>To</td><td>Professor Isabella Mendoza
Department of Environmental Engineering</td></tr>
<tr><td></td><td>From</td><td>Samantha Pearson *SP*</td></tr>
<tr><td></td><td>Subject</td><td>Progress Report on Independent Study for EnEng 4500</td></tr>
</table>

Provides initial abstract

I began work on this independent study on 20 January, after you and Professor Fuller accepted my proposal. This is the progress report scheduled in the proposal. It describes the work completed and the work remaining in the study and concludes with an appraisal of the progress of the study.

Defines project

Project Description

The purpose of this independent study is to review literature pertinent to your planned proposal for a pilot project to cleanse soils polluted by such contaminants as diesel fuel and gasoline. The review includes four major tasks:

- Search for literature concerning the causes and solutions for pipe failures during steam injection.
- Search for funding sources for the pilot project.
- Organize and analyze the information gathered.
- Write a literature review based on the information gathered.

Reports progress to date including key sources located and potential funders

Work Completed

During this reporting period, I focused on the first two tasks: finding information concerning pipe failures and funding. I found excellent information in both areas. Here, I will briefly outline what I have found:

Pipe Failures

A key article in my search for information on pipe failures was Eva L. Davis's "Steam Injection for Soil and Aquifer Remediation," in EPA's *Ground Water Issue,* January 1998. Davis points out that steam injection is a technology first used to recover oil from depleted oil fields that has been adapted for use

Page 2
Professor Mendoza
24 February 2004

in cleansing polluted soils. The problem of pipe failures was known to petroleum engineers and their solutions are still pertinent. Davis's article led me to several other excellent sources:

    C. F. Gates and B. B. Holmes, *Thermal Well Completions and Operation*, Seventh World Petroleum Congress Proceedings 3 (1967): 419–29.
    S. M. Ali Farouq and R. F. Maldau, "Current Steam Flood Technology," *J. Petroleum Technology* October 1979: 1332–42.
    M. M. Schumacher, *Enhanced Recovery of Residual and Heavy Oils*, 2nd ed. (Park Ridge, NJ: Noyes Data Corporation, 1980).
    C. Chu, "State of the Art Review of Steamflood Field Projects," *J. Petroleum Technology* October 1985: 1887–1902.

These sources recommend steel casing over PVC or fiberglass because of the extreme expansion and contraction that goes on during steam injection. They also provide mixture formulas for cement that will hold up under these extreme conditions.

Funding
Numerous agencies provide funding for polluted site remediation, but not all are interested in funding test or pilot programs. I have listed some of those who do fund such programs:

- National Environmental Technology Test Site Program, a partnership of the Environmental Protection Agency (EPA) and the Department of Defense (DoD)
- Remediation Technologies Development Forum, a partnership of the EPA, the DoD, and the Department of Energy (DOE)
- Air Force Center for Environmental Excellence
- Program Research and Development Announcements, EPA

The DoD and Air Force programs seem particularly attractive. Their past operations have led to many sites contaminated with gasoline and jet fuel. They seem quite serious about cleaning up these sites and are interested in innovative ways to accomplish the work.

Page 3
Professor Mendoza
24 February 2004

Work Remaining
The work remaining includes gathering more information. I am gathering detailed specifications for pipe construction and installation. I am contacting the funding agencies to inquire about their proposal procedures and the criteria they use in judging proposals.

When I have analyzed and organized the information, I will prepare the first draft of the review in time to get it to you and Professor Fuller by 22 March. Following the discussion with you, I will have the final report to you by no later than 5 April. I hope the review will be a well-documented summary of my information that will be useful for your proposal.

Overall Appraisal
The library and Internet search has gone well. I have copious notes that when analyzed and organized should be of great use to you. The specifications for construction and installation of steam injection pipes are quite detailed. Air Force funding seems particularly promising.

cc. Professor William Fuller

Report 2: Progress Report

Memorandum

To: Dr. Phillip Lathrop
From: Robert Hagen
Date: April 3, 2____
Subject: Progress Report for *Tae Kwon Do Student Manual*

Introduction
The title of my project is *Tae Kwon Do Student Manual*. The project is a print manual for students of Centerville Tae Kwon Do. The purpose of the manual is to provide students and their parents with a useful source of information about Centerville Tae Kwon Do. The student manual will cover all the material that students of Centerville Tae Kwon Do are required to learn from white belt to first-degree black belt. The primary audience for the manual is students of Centerville Tae Kwon Do, both at the beginner and advanced levels. The secondary audience is the instructor and parents of the Centerville Tae Kwon Do students.

Work Completed
When I conceived of the idea to write a Tae Kwon Do student manual, I had to define the need, audience, purpose, and scope. After defining those elements of the project, I began to conduct research.

On March 8 I drove to Centerville and met with Masters Tony Lopez and Victor Martinez. They gave me direction as to what would be valuable in a student manual. Master Lopez, for instance, recommended that I include a glossary of Korean terms. Master Martinez provided me with the documentation that he gives to beginner students, which helped with the writing of the student manual.

While in Centerville, I also met with Ron Avery, a personal friend who agreed to design the cover of the student manual. Ron and I discussed the tone that the cover should create and specifics about the central image and typography. We decided to use a central image of two figures kicking simultaneously encircled by a glowing globe. We also decided that the words "Tae Kwon Do" would run across the top of the globe, and the words "Student Manual" would run along the bottom of the globe.

When I arrived back in Iowa City, I drafted a detailed outline for the student manual. I also completed a written proposal and a proposed timetable for the completion of the project. I then began to write the first draft of the student manual. The writing of the first draft went

Dr. Lathrop   2   April 3, 2____

smoothly, and it is fully complete. I did not run into any complications. In fact, I saved several hours by using the Master Document feature in Microsoft Word. This was the first time that I used the Master Document feature, and I feel that it made the writing process dramatically easier. The first draft is complete.

Work Remaining
The following is a list of tasks that remain to be done to complete the student manual:

- Make revisions based on first set of peer edits and comments from you
- Meet again with Masters Lopez and Martinez and have them verify the technical accuracy of the student manual
- Meet with Ron Avery and have him add "Student Manual" to the cover
- Create a table of contents
- Conduct a usability test with three Centerville Tae Kwon Do students, two parents of Centerville Tae Kwon Do students, and one Centerville Tae Kwon Do instructor
- Complete Progress Report Two and hand in second draft
- Make further revisions based on second set of peer edits and comments from you
- Complete and hand in the final draft of the student manual

Conclusion
Overall, I am making good progress on the student manual. As I have stated earlier, the first draft is complete. By the next reporting period, I plan to have a revised draft completed. This draft will reflect the revisions suggested by you and my peers, as well as the results from a usability test, which I plan to conduct by the next reporting period. I also plan to have a revised draft of the cover from Ron Avery. I do not anticipate any complications.

Report 3: Research Report

Thawing Regimes for Freezer-Stored Container Stock

Robin Rose and Diane L. Haase

Project leader and associate director, Nursery Technology Cooperative, Oregon State University, Department of Forest Science, Corvallis, Oregon

*Three thawing regimes were applied over a 6-week period to frozen Douglas-fir (*Pseudotsuga menziesii *(Mirb.) Franco.), western larch (*Larix occidentalis *Nutt.), and ponderosa pine (*Pinus ponderosa *Dougl. ex Laws.) container stock: (1) rapid thaw followed by cold storage, (2) slow thaw, and (3) freezer storage followed by rapid thaw. Seedlings were outplanted to 3 sites in north-central Washington. A subsample of seedlings was evaluated for root growth potential (RGP) at the time of outplanting. Seedling performance was assessed after the first and second growing seasons. Although there were significant differences among species, thawing regime did not affect seedling growth or survival after 2 growing seasons nor did it affect RGP. The results indicate that seedlings can tolerate variations in thawing practices that may occur due to weather or other circumstances beyond control. However, it is noted that it may be best to keep seedlings in freezer storage for as long as possible in order to prevent storage molds.* Tree Planters' Notes 48 (1/2): 12–17; 1997.

Freezer storage of container seedlings, although an accepted practice in the nursery industry, is still a relatively misunderstood technique in some forest nurseries and reforestation organizations. Research and experience have shown that freezer storage can be a valuable management tool to a successful reforestation program. Freezer storage gives the nursery greater flexibility by allowing for lifting during late autumn and shipping the following spring. This results in a more balanced work load at the nursery and an effective "surge buffer" between nursery and field production (Hee 1987). Colombo and Cameron (1986) found that freezer storage of container black spruce—*Picea mariana* (Mill) B.S.P.—allows managers to safely delay budset of a late-sown crop, thereby reaching minimum acceptable height, without the risk of winter damage associated with outdoor storage. Furthermore, freezer storage is more suitable for periods in excess of 2 months, because carbohydrate depletion and

2   *Tree Planters' Notes*

storage molds can be a problem with long-term cold (2°F) storage (Ritchie 1982, 1984).

Freezer storage is often necessary to maintain crop dormancy when late-season planting is required in snowed-in units, especially for stock to be planted to high-elevation sites. Odlum (1992) noted that black spruce seedlings kept in frozen storage had greater subsequent root and shoot growth than those wintered outdoors, especially for those outplanted at a later date. Ritchie (1984, 1989) found that the rate of dormancy release in bareroot Douglas-fir—*Pseudotsuga menziesii* (Mirb.) Franco—seedlings was substantially retarded by freezer storage compared to those left in the nursery bed resulting in an expansion of the planting window and a higher, more uniform, physiological quality. Likewise, Lindström and Stattin (1994) found that freezer-stored seedlings of Norway spruce (*Picea abies* (L.) Karst.) and Scots pine (*Pinus sylvestris* L.) had a greater tolerance to freezing in the spring than those that were stored outdoors.

A concern with freezer storage is the thawing process. One thawing method commonly used is to allow the stock to thaw very slowly at temperatures just above freezing over a period of several weeks. Another method is to place seedlings in an area with ambient temperatures for several days prior to outplanting. The standard thawing practice for Weyerhaeuser nurseries is to spread seedling pallets out and allow them to thaw at ambient temperature (10 to 15°F) for 3 to 5 days (bare-root seedlings) and for 10 to 15 days (container seedlings) (Hee 1987). Whether thawed rapidly or slowly, field foresters prefer to have the stock thawed just prior to outplanting. However, changing weather conditions or other circumstances beyond control can result in thawed stock being held for several weeks in cold storage prior to outplant. Hee (1987) noted that it is best to plant seedlings as soon as they have thawed, but also noted that they can be held in cooler storage after thawing for up to 4 weeks without detriment.

The objective of this study was to examine the effects of 3 thawing regimes on the subsequent quality of 3 species of container-grown conifer seedlings outplanted to 3 sites. The thawing regimes were designed to simulate circumstances typically encountered with frozen stock. The null hypothesis was that there would be no differences in seedling field performance for any of the species due to thawing treatment.

Materials and Methods

Douglas-fir, western larch (*Larix occidentalis* Nutt.), and ponderosa pine (*Pinus ponderosa* Dougl.) container stock (1-year-old Styro-8) were used in this study. For each species on each outplanting site, seedlings were from the same seedlot. Seedlings were grown and freezer stored under standard nursery practices.

Seedlings were shipped frozen to the Leavenworth District of the Wenatchee National Forest in late March to early April 1995, depending on the expected date of planting for each site. Three thaw schedule treatments were applied over a 6-week period as follows:

1. Seedlings were placed under a rapid thaw (5 days at 7°C = 44.6°F) 6 weeks before expected outplanting, then held in cold storage (1°C = 33.8°F) until outplanting.
2. Seedlings were placed in cold storage for a slow thaw (6 weeks) before outplanting.
3. Seedlings were kept in freezer storage (–2°C = 28.4°F) until 1 week before outplanting, when they were placed under a rapid thaw.

Telog temperature recorders (Model 2103, Telog Instruments Inc., Victor, NY) were placed with seedlings in each thawing treatment. Because there were a limited number of Telogs available and because Telog data cannot be examined until it is downloaded to a computer, additional digital temperature probes were placed with the seedlings and monitored weekly.

Seedlings were outplanted to 3 sites on the Wenatchee and Okanogan National Forests in north-central Washington as follows:

- Twisp District, Okanogan National Forest; high-elevation (1,372 m = 4,500 ft) dry site. The slope is 10 to 40% with a northeastern aspect, with light slash and vegetation. All 3 species were planted on June 1, 1995.
- Leavenworth District, Wenatchee National Forest; low-elevation (610 m = 2,000 ft) dry site in area burned by 1994 wildfire. Annual precipitation is 53 to 76 cm (20 to 30 in). Soil is sandy to clay loam.

4 *Tree Planters' Notes*

The slope is 60% and the burned trees (avg. dbh = 10 cm = 4 in) were left standing. Douglas-fir and ponderosa pine were planted on April 20, 1995.

- Naches District, Wenatchee National Forest; high-elevation (1,219 m = 4,000 ft) temperate site. The slope is 15% with a western aspect. Douglas-fir and western larch were planted on May 31, 1995.

Seedlings were outplanted at about the same time that the site was scheduled to be operationally planted. Because of late-winter conditions, the 6-week thawing period was extended by 7 to 10 days for seedlings planted on the Twisp and Naches Districts. For each site, all seedlings were planted on the same day. Seedlings were planted at a spacing of 1.5 ˇ 1.5 m (= 4.9 ˇ 4.9 ft).

Initial height and survival were measured and recorded 2 weeks after outplanting and again at the end of the first and second growing seasons (September 1995 and August 1996). In addition, a damage/vigor assessment (incidence of browse, chlorosis, etc.) was recorded for each seedling.

In addition to the outplanted seedlings, a subsample of 15 seedlings of each species/treatment from the Leavenworth and Twisp sites were sent to International Paper's Lebanon facility shortly after seedlings were outplanted (that is, after treatment) and evaluated for root growth potential. These seedlings were potted and allowed to grow in a greenhouse for 3 weeks, then evaluated for the number of seedlings with new roots.

The experimental design consisted of a split-plot design with 5 blocks, 2 or 3 species per site (whole plots), 3 thaw treatments (subplots), and 10 seedlings in each block/species/treatment for a total of 450 seedlings on the Twisp site and 300 seedlings on the Leavenworth and Naches sites. All seedlings were labeled and randomly planted within a block.

An analysis of variance (ANOVA) was performed on all data to determine if thaw treatment has a significant effect on subsequent seedling performance. Differences among mean values for species and treatment were determined using Fisher's protected least significant difference procedure. Statistical Analysis Software (SAS Institute 1989) was used for all data analyses.

Results

It took about 5 days to accomplish the rapid thaw (treatments 1 and 3) and about 3 weeks for the slow thaw (treatment 2) (figure 1).

As would be expected, there were significant differences in field performance among species on each site (figures 2 and 3). However, there did not appear to be any meaningful differences among thawing treatments. During the first season, there were significant treatment by species interactions for both height and growth on the Leavenworth and Naches sites (figure 2). However, despite the statistical significance between treatments, the differences in first-year average height and growth may not be significant from a reforestation perspective, as the differences are small (1 to 3 cm = .4 to 1.2 in) and the ranking does not follow any pattern with regard to the treatments. For example, treatment 1 Douglas-

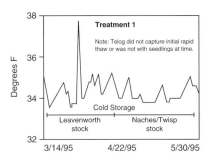
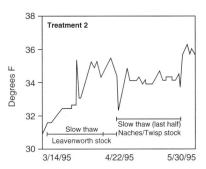
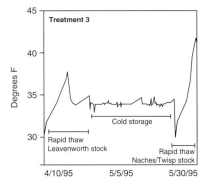

Figure 1—*Output from Telog temperature recorders showing the thawing process of each treatment.*

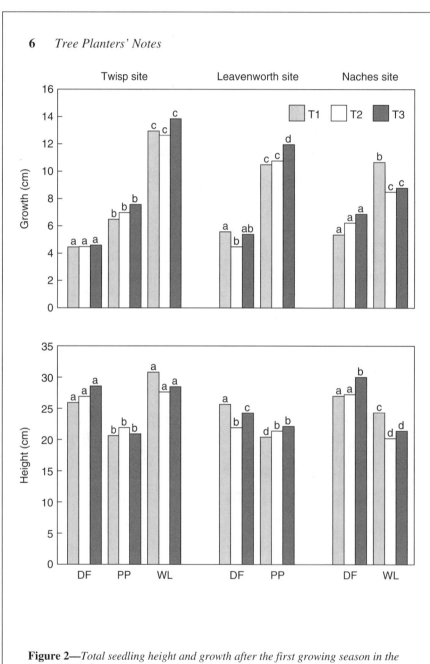

Figure 2—*Total seedling height and growth after the first growing season in the field (1995). On each site, bars with different letters are significantly different at the $\propto \leq 0.05$ level.*

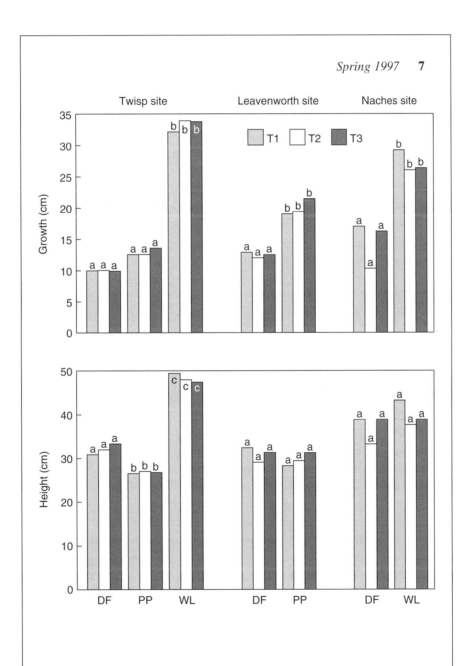

Figure 3—*Total height and growth after two growing seasons (1996). Although species differed significantly, there were no significant differences between thawing treatments. On each site, bars with different letters are significantly different at the $\propto \leq 0.05$ level.*

8 *Tree Planters' Notes*

fir had the greatest height on the Leavenworth site, whereas treatment 3 Douglas-fir had the greatest height on the Naches site. Similarly, treatment 3 ponderosa pine had the most growth on the Leavenworth site whereas treatment 1 western larch had the most growth on the Naches site. During the second growing season, there were no significant differences among thawing treatments for total height, seasonal height growth, or total height growth on any of the 3 sites (figure 3).

Survival averaged 77% on the Twisp site and 96% on the Leavenworth site regardless of species or treatment. On the Naches site, survival was not influenced by treatment but was very poor for Douglas-fir (23%) compared to western larch (83%). Thaw treatment had no effect on root growth potential.

Discussion

We found that thawing regime did not affect subsequent seedling field performance. In a similar study, Camm and others (1995) reported that there were very few differences between white spruce (*Picea glauca* (Moench) Voss) and Engelmann spruce (*Picea engelmannii* Parry) seedlings planted either directly from the freezer or after 9 days of thawing. The latter broke bud 3.3 days earlier than those planted directly from the freezer but had a less uniform budbreak. Height, shoot, and root mass did not differ after 3 months of growth. Camm and others (1995) suggest that a suitable on-site operational protocol for rapid thawing might be to lay frozen bundles on the ground at ambient temperature overnight. Additional possible benefits to this approach that they mention include reductions in handling costs, secondary storage facilities, and losses caused by refrigerator failure (Camm and others 1995).

The idea of a long, slow thaw has been to allow normal physiological processes to fully resume prior to planting. However, this may not be necessary because recovery of water potential after thawing spruce seedlings took hours, not days, once ice crystals left the roots (Camm and others 1995). As a result, these authors recommend against the practice of slowly thawing seedlings for up to several weeks before shipping to the plantation site because fungi (*Botrytis* spp.) often proliferate on seedlings held above freezing in the dark for extended periods. Another study showed that steady-state respiration rates increase significantly during thawing and hence have the potential to greatly deplete carbohydrate reserves, especially over time (Lévesque and Guy 1994).

On the other hand, Odlum (1992) stated that rapid thawing of stock can result in damage or mortality attributable to shoots rapidly rising to ambient thaw temperature, while seedling plugs remain frozen, due to their higher water content. Thus, foliar transpiration without water availability from the roots results in desiccation. Odlum recommended that stock be thawed slowly as described by Koistra and others (1989); seedlings are first exposed to 5°C until completely thawed. Our findings do not suggest the need for this.

Conclusions

Despite assertions in the literature of damage to seedlings caused by either rapid or slow thawing, the results of our study indicate that container seedlings can withstand variations in thawing regimes, as we described, without any detrimental effect to their subsequent field performance. However, managers concerned with post-storage fungal infection should consider using short thawing intervals.

Literature Cited

Camm EL, Guy RD, Kubien DS, Goetze DC, Silim SN, Burton PJ. 1995. Physiological recovery of freezer-stored white and Engelmann spruce seedlings planted following different thawing regimes. New Forests 10: 55–77.

Colombo SJ, Cameron SC. 1986. Assessing readiness of black spruce container seedlings for frozen storage using electrical impedance, oscilloscope square wave and frost hardiness techniques. For. Res. Note 42. Location: Ontario Ministry of Natural Resources, Ontario Tree Improvement and Forest Biomass Institute: 6 p.

Hee SM. 1987. Freezer storage practices at Weyerhaeuser nurseries. Tree Planters' Notes 38(2): 7–10.

Koistra C, Ostafew S, Lukinuk I. 1989. Cold storage guidelines. Victoria, BC: British Columbia Ministry of Forests.

10 *Tree Planters' Notes*

Lévesque F, Guy RD. 1994. Changes in respiration rates of white spruce and lodgepole pine seedlings during freezer storage and thawing and relationship to carbohydrate depletion. Poster presented at Annual Meeting of the American Society of Plant Physiologists. Portland, OR. July 30–August 3, 1994.

Lindström A, Stattin E. 1994. Root freezing tolerance and vitality of Norway spruce and Scots pine seedlings: influence of storage duration, storage temperature, and prestorage root freezing. Canadian Journal of Forest Research 24: 2477–2484.

Odlum KD. 1992. Maybe or maybe nots of frozen storage. Paper presented at OTSGA container workshop, 1991 October 1–3; Kirkland Lake, ON. Location: Ontario Tree Seedling Growers Association and Ontario Ministry of Natural Resources.

Ritchie CA. 1982. Carbohydrate reserves and root growth potential in Douglas-fir seedlings before and after cold storage. Canadian Journal of Forest Research 12: 905–912.

Ritchie GA. 1984. Effect of freezer storage on bud dormancy release in Douglas-fir seedlings. Canadian Journal of Forest Research 14(2): 186–190.

Ritchie GA. 1989. Integrated growing schedules for achieving physiological uniformity in coniferous planting stock. Forestry 62 (Suppl.): 213–227.

SAS Institute. 1989. SAS/STAT User's Guide, Version 6, Fourth ed. Cary, NC: SAS Institute. 846 p.

Report 4: Report of Findings

Environmental Consultants
2063 Peach Tree Street
Suite 260
Atlanta, GA 30747
Phone: 404-555-1940
Fax: 404-555-9003
E-mail: nlarsen@enco.com

November 20, 2005

Mr. James Morris
Chief Executive Officer
Albany Office Products
22 Oglethorpe Road
Albany, GA 30278

Subject: Inspection Findings and Recommendations

Dear Mr. Morris:

As you engaged us to do, we have examined the environmental problems in your company headquarters. This report provides background for our report, states the problems our inspection uncovered, and recommends solutions for them.

Background

For the last year, the workers in your two-story office building have experienced higher-than-average health problems. They have suffered from watery eyes, nasal congestion, coughing, difficulty breathing, headache, and fatigue. Last winter, your workers had a high incidence of flu, and lost work days grew to unacceptable levels because of it. During the last six months, three of your employees with asthma had to be hospitalized.

Page 2
Mr. James Morris
November 20, 2005

Although these problems were not limited to your first-floor offices, they were more prevalent there than on the second floor. All these reactions pointed to the presence of excessive biological pollution in your building, particularly on the first floor.

Biological pollutants are found everywhere. Molds, bacteria, and viruses are commonly found in office buildings such as yours. People exposed to such pollutants may suffer allergic reactions, infections, and even serious toxic reactions in the central nervous system and the immune system.

Biological pollutants need moisture to grow and spread. When moisture levels in a building are lowered, the level of pollutants and the reactions to them are greatly reduced. Therefore, our inspection of your building focused on moisture problems, particularly on the first floor.

Findings of the Inspection

Our inspection found major problems with your first-floor carpet and heating and air-conditioning ducts. We found also a minor problem with the large number of coffee makers in your building.

First-Floor Carpeting

Your building is slab constructed, and bare concrete underlies all of the matting and wall-to-wall carpet on the first floor. Moisture has passed through the concrete and allowed mold to grow and spread in and under the matting and the carpet. Spot inspections indicate that more than 80 percent of the first-floor carpet has mold growing underneath it. Where mold is found, you can be sure that bacteria and viruses also flourish. This condition most certainly explains the high incidence of health problems on the first floor.

Heating and Air-Conditioning Ducts

The heating and air-conditioning ducts are full of dust and are beginning to show signs of mold. The heater and air conditioner had not been properly

Page 3
Mr. James Morris
November 20, 2005

cleaned, and the system filters were clogged with dust. These conditions explain the incidence of health problems on the second floor.

Coffee Makers

Office policy obviously does not regulate the use of coffee makers throughout the building. We found 18 coffee makers plugged in and working during our inspection. These coffee makers put a great deal of moisture into the air, encouraging the growth of pollutants.

Recommended Solutions

The problems encountered in your building all have ready solutions.

First-Floor Carpeting

The first-floor carpeting and matting are too far gone to be salvaged. They must be taken up and discarded. The concrete floor must be professionally cleaned and disinfected. Following that, you have two alternatives:

- Lay a plastic vapor barrier on the concrete and cover that with a subfloor of insulation and plywood. Matting and wall-to-wall carpet can then be laid on the plywood.

- A somewhat less expensive alternative would be to lay good-quality asphalt or vinyl tile on the concrete and use area rugs where carpeting is wanted.

Heating and Air-Conditioning Ducts

You need to take three steps to keep your heating and air-conditioning ducts free of pollution:

- Have the heating and air-conditioning units and their ducts professionally cleaned as soon as possible but not until after the carpet

Page 4
Mr. James Morris
November 20, 2005

problem has been resolved. (Removing the carpet, cleaning the concrete floor, and laying subflooring or tile will kick up dust and dirt.)

- Contract with heating and air-conditioning professionals to have them clean your heaters and air conditioners at the start of each heating and cooling period.

- Arrange to have your building cleaners change your system filters monthly.

Coffee Makers

Remove the coffee makers from the various offices. If you wish, in some well-ventilated space, place one or two larger coffee makers that everyone can use.

The recommended solutions follow the guidelines laid down by the U.S. Consumer Product Safety Commission and the American Lung Association. While air pollution cannot be completely eliminated, short of extraordinary measures, carrying out the work recommended will restore a healthful environment to your workplace.

If you want our assistance in locating the professionals to carry out the needed work, let us know. Thank you for letting us help you.

Sincerely,

Nancy Larsen

Nancy Larsen
Chief of Inspections

NL: siu

Report 5: Laboratory Report

Here is an example of design that could be improved for clarity. The designer has placed two addresses plus basic report information in the same area, using the same font. Because these different kinds of information are presented in the same way, the reader might miss the basic report information in the upper-right corner, assuming that it's just another address. The designer should consider using different fonts, or different placement entirely.

Wisely, the writer has placed the reason for the report at the beginning of the document. Readers should not wonder why they are reading a document. If they do, they're not likely to continue.

Again, the writer makes a wise decision by putting the testing methods at the beginning. Keep in mind that the laboratory report is a specific genre audiences will recognize. For this reason certain elements are expected in a lab report. They usually include *a summary of the experiment, a presentation of the results,* and *a discussion of the results.* If any of these elements are missing, the audience could become confused. The purpose of the document will be lost.

FEND-LAB, INC.
2314 Universal St., Suite 192
San Francisco, CA 94106
(325) 555-1327
www.fendlabcal.com

Test Address
NewGen Information Technology, LLC
3910 S. Randolph
Slater, CA 93492

Client
Brian Wilson
Phone: 650-555-1182
Fax: 650-555-2319
e-mail: brian_wilson@cssf.edu

Mold Analysis Report
Report Number: 818237-28
Date of Sampling: 091204
Arrival Date: 091404
Analysis Date: 091904
Technician: Alice Valles

Lab Report: Mold Test

In this report, we present the results of our testing for mold at the offices of NewGen Information Technology, at 3910 S. Randolph in Slater, California. Our results show above-normal amounts of allergenic mold, which may lead to allergic reactions among the residents.

Testing Methods

On 12 September 2004, we took samples from the test site with two common methods: Lift Tape Sampling and Bulk Physical Sampling.

Lift Tape Sampling. We located 10 areas around the building where we suspected mold or spores might exist (e.g., water stains, dusty areas, damp areas). Using 8-cm-wide strips of transparent tape, we lifted samples and pressed them into the nutrient agar in petri dishes. Each sample was sealed and sent to our laboratory, where it was allowed to grow for one week.

This document maintains simplicity in its design. There are only four section headings, which makes it easy to scan, and thus easy to navigate. Readers can locate information quickly. They expect to read about what was found and how it affects them, and the document's design meets those expectations.

This graphic is effective in supporting the document's purpose, which is to communicate what kinds of mold have grown and to what extent. There are many ways in which this information could be presented, but because these authors/designers understand the document's purpose and their audience's needs, the essential information is displayed in a clear, concise format.

Bulk Physical Sampling. We located 5 additional areas where we observed significant mold growth in ducts or on walls. Using a sterilized scraper, we removed samples from these areas and preserved them in plastic bags. In one place, we cut a 1-inch-square sample from carpet padding because it was damp and contained mold. This sample was saved in a plastic bag. All the samples were sent to our laboratory.

At the laboratory, the samples were examined through a microscope. We also collected spores in a vacuum chamber. Mold species and spores were identified.

Results of Microscopic Examination

The following chart lists the results of the microscope examination:

Mold Found	Location	Amount
Trichoderma	Break room counter	Normal growth
Geotrichum	Corner, second floor	Normal growth
Cladosporium	Air ducts	Heavy growth
Penicillium spores	Corkboard in bathroom	Normal growth

Descriptions of molds found:

Trichoderma: Trichoderma is typically found in moistened paper and unglazed ceramics. This mold is mildly allergenic in some humans, and it can create antibiotics that are harmful to plants.

Geotrichum: Geotrichum is a natural part of our environment, but it can be mildly allergenic. It is usually found in soil in potted plants and on wet textiles.

Cladosporium: Cladosporium can cause serious asthma and it can lead to edema and bronchiospasms. In chronic cases, this mold can lead to pulmonary emphysema.

Penicillium: Penicillium is not toxic to most humans in normal amounts. It is regularly found in buildings and likely poses no threat.

This paragraph displays *unity*, which is essential in technical writing. Each sentence supports the topic sentence of the paragraph. The point is to meet readers' expectations. In general, when you write a paragraph, the first sentence should state the subject, and the sentences that follow should discuss it.

Discussion of Results

It does not surprise us that the client and her employees are experiencing mild asthma attacks in their office, as well as allergic reactions. The amount of Cladosporium, a common culprit behind mold-caused asthma, is well above average. More than likely, this mold has spread throughout the duct system of the building, meaning there are probably no places where employees can avoid coming into contact with this mold and its spores.

The other molds found in the building could be causing some of the employees' allergic reactions, but it is less likely. Even at normal amounts, Geotrichum can cause irritation to people prone to mold allergies. Likewise, Trichoderma could cause problems, but it would not cause the kinds of allergic reactions the client reports. Penicillium in the amounts found would not be a problem.

The results of our analysis lead us to believe that the Cladosporium is the main problem in the building.

Conclusions

The mold problem in this building will not go away over time. Cladosporium has obviously found a comfortable place in the air ducts of the building. It will continue to live there and send out spores until it is removed.

We suggest further testing to confirm our findings and measure the extent of the mold problem in the building. If our findings are confirmed, the building will not be safely habitable until a professional mold remover is hired to eradicate the mold.

Ignoring the problem would not be wise. At this point, the residents are experiencing mild asthma attacks and occasional allergic reactions. These symptoms will only grow worse over time, leading to potentially life-threatening situations.

Contact us at (325) 555-1327 if you would like us to further explain our methods and/or results.

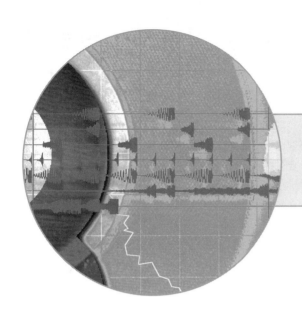

Oral Presentations and PowerPoints

Public speaking is an important—and for many people, frightening—task. It requires some degree of confidence and skill, but new technology can help provide a structure for public presentations, whether they are for a small group of colleagues, a potential client, or a large audience. As with virtually all forms of business and technical communication, knowing your audience is paramount: Who are the people you are talking to? Why are they here? What do they want to learn from you? What will appeal to them—and give them the impetus to do what you want them to do?

Using visual aids like slides or PowerPoints can help you organize your presentation, give you confidence while you talk, and provide a printed handout for your audience to save them from having to take extensive notes. But visual aids are just that—aids. They cannot do the central work of an oral presentation. Being prepared—understanding your topic, understanding your audience, and understanding the purpose of your presentation is the most important aspect of the task.

Quick Reminders for Preparing a Slide or PowerPoint Presentation

- **Use visuals to support important points—not to provide the text of your presentation.** Never put more than 6 lines on a slide (three is better) and always use bullets rather than narrative text. If you put whole paragraphs on the screen, people will stop listening and start reading.

- **Think about design.** Do you want your background to be neutral, so it doesn't distract from the text? Do you want to use repetitive graphics for design purposes only, or content-related graphics? What is the relationship between design and content?

- **Keep it simple.** Use big type that can be easily read from the back of the room, a simple background that doesn't conflict with the text color you are using, and stay away from sounds or dissolves unless they are meaningful rather than distracting.

PowerPoint 1

DNA Fingerprinting

It will never be useful in the courtroom unless juries can understand it.

Three parts are necessary to understand DNA Fingerprinting

- What is DNA?
- What does a lab do with DNA?
- Can someone else have the same fingerprint?

What is DNA?

- It is found in all living cells.
- Contains code
- Structure is a double helix-twisted ladder

What is DNA?

- Consists of 4 "bases"---A,T,G,C
- Bases are complimentary
- This helix can be separated (unzipped)

```
G A T A C G G A T C
C T A T G C C T A G

G A T A C G G A T C
C T A T G C C T A G
```

Two more facts about DNA that make it work as identification

- Of the 3 billion or so base pairs comprising a DNA molecule, only 3% are used to store info---
- The remaining "intergenic" sequence is very repetitive but distinctive.

The DNA fingerprinting process

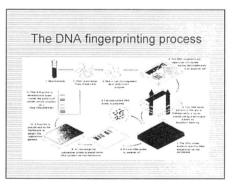

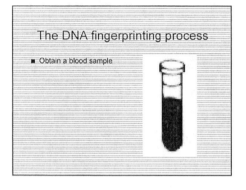
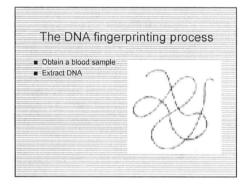
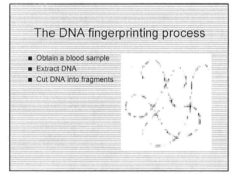
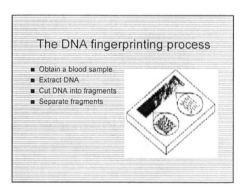
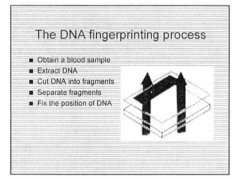
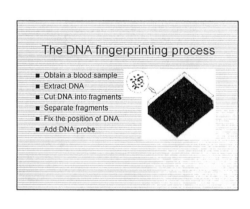

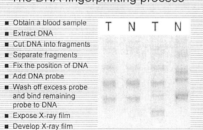

Part 2: Scenarios and Case Studies

Introduction

A scenario is typically a fictionalized situation—although sometimes based in reality—describing a set of circumstances and asking for a response of some kind. Case studies, which are more extended scenarios, are more often based on real circumstances. In both cases, you will probably be presented with a description of the circumstances and people, possibly some sample documents, and a request that you make a decision or find a solution. Sometimes you will be asked to play a specific role—to put yourself into the scene and respond from that perspective.

As you read the following case studies and scenarios, note details, character traits, and circumstances that you think might help you respond. What are the key words? What are the characteristics of the players? Who will make up the audience for whatever response you create? What do they want or need from you? What kind of document or documents might be written in response to the scenario?

Case Study 1: A Visit to the Forbidden City

A Visit to the Forbidden City:
A Sign of the Times

SAM DRAGGA
Texas Tech University

Abstract: On a visit to Beijing, China, a technical communicator offers to revise the English-language version of a sign at the Forbidden City that explains the significance of the historic site to foreign tourists. Initially, the job seems simple enough as he focuses on meeting the information needs of the designated audience. He soon realizes, however, that his revision ignores the genuine complexities of Chinese culture.

Note: The following case is fictitious. However, Figure 1 replicates a real sign previously on public display at the Forbidden City, and Figure 3 replicates a real sign exhibited today at this historic site.

Background for the Case

Thomas Wild is chief executive officer of Wild Solutions, Inc., in Seattle, Washington. Wild Solutions is a small company that employs five writers, a supervisor/editor, and two artists/designers. Wild started the company in 1990, after he graduated from the University of Washington with a bachelor's degree in psychology and a master's degree in technical communication. The company specializes in devising communication solutions for business organizations. More and more of the clients are multinational corporations.

Wild does a little international traveling, chiefly in Asia, soliciting business for his company and meeting with existing clients. He has been to India, Japan, and once to the People's Republic of China, specifically Shanghai.

| 2 | *A Visit to the Forbidden City* |

In 1996, he visits Beijing. While in the capital city, at the urging of his client, Gu Baohui of Xinghua Computer Corporation, Wild decides to delay their business meeting a day while he visits a major attraction for millions of foreign tourists—the Forbidden City, once the imperial palace of the emperors of China. He enjoys his visit to this historic site, but like a good technical communicator, he notices that the sign explaining the museum to foreign visitors is defective and could be revised. On the left side of the gate to the palace is a wooden sign with painted Chinese characters. On the right side is a wooden sign with the painted English translation (see Figure 1). At least Wild assumes it is a translation: although he speaks a little Chinese (chiefly common greetings), Wild has a quite limited understanding of Chinese characters and has hired a local interpreter for his business negotiations.

Aside from the grammatical and spelling errors, Wild notices immediately that the English translation displays several hyphenation errors. Thus, at the end of a line, instead of dividing words between syllables, the sign divides words arbitrarily (e.g., "re/sidence" and "pala/ce"). In addition, the use of all uppercase letters makes the sign difficult to read. And the sign offers a lot of information that typical tourists might consider unnecessary, such as both the Chinese and English names for each of the buildings.

At a meeting with his client on the following day, Wild mentions how delighted he was with his visit to the Forbidden City. Mr. Gu is obviously pleased. In his limited English, he reveals that his sister, Zhou Peiya, is a supervisor at the museum. Mr. Gu also explains that the museum exhibits only a fraction of the available imperial treasures and, according to his sister, will soon be receiving major new funding from private and public sources. Wild confesses that he thought the English language signs could be revised and proposes that his company would be willing to do the writing as a donation to the museum. Mr. Gu declares that he will notify his sister of Wild's generous offer.

Revising the Sign

Later that afternoon, Ms. Zhou telephones Wild at his hotel and invites him to a meeting the next morning at the Forbidden City. On arrival at the palace, Wild receives a private guided tour of the facility from Ms. Zhou. He is ushered through buildings typically closed to the public—rooms of furniture, clothing, jewelry, porcelain, and mechanical toys. After the tour, Wild is guided to a lovely meeting room filled with exquisite paintings and embroideries. He is introduced to several museum officials and served tea. "The new sign must help tourists appreciate the treasures of Chinese civilization." Ms. Zhou explains. "Millions of people come every year. We have audiotapes in several languages and players. . . ."

Yes, I did the audio tour yesterday, Wild interrupts. "It's really good. Quite informative!"

Ms. Zhou smiles. "It was all that was possible with so little money."

"Oh, is that right?" commiserates Wild.

Ms. Zhou looks at the museum officials, looks at Wild, and smiles again. She proceeds, "We have domestic tourists and foreign tourists. The majority of the

A Visit to the Forbidden City   3

THE PALACE MUSEUM

THE IMPERIAL PALACE, POPULARLY KNOWN AS THE FORBIDDEN CITY, WAS THE PERMANENT RE-SIDENCE OF THE EMPERORS OF THE MING AND QING DYNASTIES. BUILT IN 1406–1420, THE IMPERIAL PALACE HAS A HISTORY OF 560 YEARS. OCCUPYING AN AREA OF 720,000 SQUARE METRES WITH OVER 9000 ROOMS THE IMPERIAL PALACE IS THE LARGEST AND MOST COMPLETE GROUP OF ANCIENT BUILDINGS WHICH CHINA HAS PRE-SERVED TO THE PRESENT. IN 1961, THE IMPERIAL PALACE WAS LISTED BY THE STATE COUNCIL AS ONE OF THE IMPORTANT HISTORICAL MONUMENTS UNDER THE PROTECTION OF THE GOVERNMENT.

THE BUILDINGS OF THE IMPERIAL PALACE ARE DIVIDED INTO TWO PARTS. THE FRONT PART, OR THE OUTER COURT, CONSISTS OF THE THREE GREAT HALLS—TAI HE DIAN (HALL OF SUPREME HARMONY), ZHONG HE DIAN (HALL OF MIDDLE HARMONY), AND BAD HE DIAN (HALL OF PRESERVING HARMONY), WHERE THE EMPEROR HELD IMPORTANT CEREMONIES. THE REAR PART, OR THE INNER COURT, CONSISTS OF THE THREE REAR PALACES—QIAN QING GONG (PALACE OF HEAVENLY PURITY), JIAO TAI DIAN (HALL OF UNIION), KUN NING GONG (PALACE OF EARTHLY TRANQUILITY), YANG XIN DIAN (HALL OF MENTAL CULTIVATION), PLUS THE SIX EAST PALACES, THE SIX WEST PALACES AND YU HUA YUAN (IMPERIAL GARDEN), WHERE THE EMPEROR HANDLED ROUTINE AFFAIRS AND WHERE THE EMPEROR AND HIS EMPRESS AND CONCUBINES LIVED AND SPENT THEIR LEISURE HOURS. FROM MING TO QING DYNASTIES, A TOTAL OF 24 (14 MING AND 10 QING) EMPERORS LIVED HERE AND EXERCISED SUPREME FEUDAL AUTOCRATIC POWER OVER THE COUNTRY. THE QING DYNASTY WAS OVERTHROWN IN THE REVOLUTION OF 1911. IN 1914, THE MUSEUM OF ANTIQUITIES WAS HOUSED IN THE THREE GREAT HALLS OF OUTER COURT. IN 1925, THE WHOLE REAR PART OF THE PALACE WAS TURNED INTO THE PALACE MUSEUM. IN 1947, THE MUSEUM OF ANTIQUITIES WAS MERGED INTO THE PALACE MUSEUM.

SINCE THE FOUNDING OF THE PEOPLE'S REPUB- LIC OF CHINA IN 1949, NOT ONLY THE ANCIENT BUILDINGS HAVE BEEN REPAIRED, BUT ALSO A LARGE AMOUNT OF WORK IN THE ACQUISITION, AR-RANGEMENT, RESTORATION AND EXHIBITION OF CU-LTURAL RELICS HAS BEEN MADE. AT PRESENT, FOR DISPLAY SOME OF THE HALLS ARE KEPT AS THEY WERE ORIGINALLY FURNISHED, AND OTHERS ARE USED TO EXHIBIT SPECIAL COLLECTIONS: ANCIENT PAINTINGS, BRONZES, CERAMICS, ARTS AND CRAFTS, JEWELRY, CLOCKS AND WATCHES, ETC., WHICH SHOW THE SPLENDID ACHIEVEMENTS OF TRADITIONAL CHINESE CULTURE.

FIGURE 1 Original Sign at the Palace Museum

4   *A Visit to the Forbidden City*

foreign tourists will speak or read English, so we have only two signs: Chinese for the Chinese and Japanese, and English for the Americans, Australians, Indians, Africans, and Europeans who visit. So the English sign is important," Ms. Zhou explains. "The people reading the English sign usually have little understanding of China."

"I will write you a good sign." Wild promises.

He exchanges business cards with the museum officials, adopting the Chinese practice of offering and receiving the cards with two hands.

On returning to Seattle, Wild immediately starts to read about the Forbidden City, quickly finishing several books, and to write a new sign. After a couple of weeks of sporadic writing and editing six different drafts, he produces a version he considers satisfactory (see Figure 2).

He e-mails his version of the new sign to Ms. Zhou.

She replies quickly. "The new sign arrived today. My colleagues and I are pleased to receive it. We appreciate your generous gift."

Two weeks later, Wild e-mails again to ask Ms. Zhou if she has questions about the new sign.

"It is not so long," she notes.

The Palace Museum

Built in 1406–1420, the Imperial Palace, popularly known as the Forbidden City, was the permanent residence of the emperors of the Ming and Qing dynasties. Its buildings are divided into two parts. The front part, or the "outer court," consists of the Hall of Supreme Harmony, Hall of Middle Harmony, Hall of Preserving Harmony, which are taken as its main body. The rear part, or the "inner court," consists of the Palace of Heavenly Purity, Hall of Union, Palace of Earthly Tranquility, Hall of Mental Cultivation plus the six east palaces, the six west palaces and the Imperial Garden, where the emperor handled routine affairs and he with his empress and concubines lived or spent their leisure hours. A total of 24 Ming and Qing emperors lived here.

Since the founding of the People's Republic of China in 1949, not only the palace buildings have been repaired, but also a vast amount of work has been performed on the arrangement, restoration, collection and exhibition of precious cultural relics. Today, some of the halls or palaces are kept as they were originally furnished, and the others are used to exhibit special art treasures of the Chinese nation, such as jewelry, ancient paintings, bronzes, ceramics, and clocks and watches.

The Imperial Palace is the largest and most complete group of ancient buildings that China has preserved to the present. It embodies the fine tradition and national style of ancient Chinese architectural art.

FIGURE 2 Proposed Revision to the Sign

A Visit to the Forbidden City   5

"Yes," Wild explains. "I focused on the essential information. That will make hurried tourists more likely to read the sign."

She also questions the use of lowercase letters.

"It's easier to read this way," Wild explains. "The similar size and shape of uppercase letters makes it difficult for readers to distinguish one letter from the next."

"We'll consider this important change," Ms. Zhou replies.

"I am so pleased you like it," answers Wild. "I was working on the revision day and night and keeping my usual clients waiting."

Wild is initially pleased with his effort. But the newly hired editor at Wild Solutions, Sylvia Jiang, a technical communication graduate from Miami University and the granddaughter of Chinese immigrants, thinks that Wild's version sacrifices a clear sense of the culture for readability. "You've eliminated the Chinese voice, the Chinese feeling of the original sign," she claims. "It's more American and less Chinese. I don't think they'll like it."

Wild is encouraged, however, because three months later, Ms. Zhou e-mails again to notify Wild that new signs will soon be installed at the Forbidden City.

On a business trip to Beijing later that year, Wild stops again at the Forbidden City. His visit is chiefly to see the new sign, but he is also meeting with Ms. Zhou to raise the possibility of his writing additional materials for the museum—brochures, funding proposals, and fliers for special exhibits.

On seeing the sign, Wild is simultaneously disappointed and irritated. The sign is substantially as he revised it, but previously deleted information has been reinserted and Chinese names have been substituted for the English names of buildings. It also uses all uppercase letters and includes obvious errors in spelling, punctuation, and grammar (see Figure 3). And though the original sign was painted wood, the new sign is engraved bronze; it is unlikely to be revised again soon. He sighs.

Wild has a meeting with Ms. Zhou in five minutes. He is unsure of how to approach this meeting. He thought the museum officials liked his revised version of the sign, but obviously he was wrong.

| 6 | *A Visit to the Forbidden City* |

> THE PALACE MUSEUM
>
> BUILT IN 1406–1420, THE IMPERIAL PALACE, POPULARLY KNOWN AS THE FORBIDDEN CITY, WAS THE PERMANENT RESIDENCE OF THE EMPERORS OF THE MING AND QING DYNASTIES. ITS BUILDINGS ARE DIVIDED INTO TWO PARTS. THE FRONT PART, OR THE "OUTER COURT", CONSISTS OF TAI HE DIAN HALL, ZHONG HE DIAN HALL AND BAO HE DIAN HALL, WHICH ARE TAKEN AS ITS MAIN BODY, PLUS WEN HUA DIAN HALL AND WU YING DIAN HALL, WHICH ARE TAKEN AS ITS TWO WINGS, WHERE THE EMPEROR HELD IMPORTANT CEREMONIES. THE REAR PART, OR THE "INNER COURT", CONSISTS OF QIAN QING GONG PALACE, JAO TAI DIAN HALL, KUN NING GONG PALACE, YANG XIN DIAN HALL PLUS THE SIX EAST PALACES, THE SIX WEST PALACES AND YUHUA YUAN GARDEN, WHERE THE EMPEROR HANDLED ROUTINE AFFAIRS AND HE WITH HIS EMPRESS AND CONCUBINES LIVED OR SPENT THEIR LEISURE HOURS.
>
> THE IMPERIAL PLACE IS THE LARGEST AND MOST COMPLETE GROUP OF ANCIENT BUILDINGS WHICH CHINA HAS PRESERVED TO THE PRESENT. IT EMBODIES THE FINE TRADITION AND NATIONAL STYLE OF ANCIENT CHINESE ARCHITECTURAL ART. IN 1961 THE IMPERIAL PALACE WAS LISTED BY THE STATE COUNCIL AS ONE OF "THE IMPORTANT HISTORICAL MONUMENTS UNDER THE PROTECTION OF THE GOVERNMENT", AND, IN 1987, IT WAS AFFIRMED BY THE UNESCO AS "THE WORLD HERITAGE".
>
> FROM MING TO QING DYNASTIES, A TOTAL OF 24 EMPERORS LIVED HERE. THE QING DYNASTY WAS OVERTHROWN IN THE REVOLUTION OF 1911 FROM THEN ON, AS THE FEUDAL IMPERIAL PALACE, THE FORBIDDEN CITY COMPLETED ITS HISTORICAL MISSION. IN 1914, THE MUSEUM OF ANTIQUITIES WAS HOUSED IN THE OUTER COURT. IN 1925, THE PALACE MUSEUM WAS ESTABLISHED.
>
> SINCE THE FOUNDING OF THE PEOPLE'S REPUBLIC OF CHINA IN 1949 NOT ONLY THE PALACE BUILDINGS HAVE BEEN REPAIRED, BUT ALSO A VAST AMOUNT OF WORK ON THE ARRANGEMENT, RESTORATION, COLLECTION AND EXHIBITION OF PRECIOUS CULTURAL RELICS HAS BEEN MADE. TODAY, SOME OF THE HALLS OR PALACES ARE KEPT AS THEY WERE ORIGINALLY FURNISHED, THE OTHERS ARE USED TO EXHIBIT SPECIAL ART TREASURES, SUCH AS JEWELLERY, ANCIENT PAINTINGS, BRONZES, CERAMICS, ARTS AND CRAFTS, CLOCKS AND WATCHES, ETC, WHICH SHOW THE AGE-OLD AND SPLENDID HISTORICAL CIVILIZATION OF THE CHINESE NATION.

FIGURE 3 Revised Sign

Case Study 2: The Heated Sidewalk Problem

The Heated Sidewalk Problem

CONSIDERATIONS

As a technical writer, you are often the person who writes the first documents for "general consumption," or outside audiences. In this situation, you have reservations about how your company is using a product in a particular situation. While reading, keep the following questions in mind:

- What are the major concerns of your company and of the college purchasing your product?
- As a new employee, what are the potential impacts on your ability to work with your colleagues on future projects?
- What information are you provided, and how reliable is that information?

This is a pretty big day for you. You've been employed as a technical writer for just six months at Michaels and Greenwall Associates, an engineering and architectural firm located in Madison, Wisconsin. You have enjoyed the varying writing jobs you've tackled, including reports, proposals, one grant, and even some public relations material. Today, for the first time, you meet with a large group of company engineers to begin gathering information for marketing a new product.

You stop by your office, drop off your coat and briefcase, gather your notepad and a few pencils, and walk down to the conference room for your 9:00 a.m. meeting with the engineers.

"Welcome. Sit down," says the lead development engineer Robert Cruzner as you walk into the room. You settle in, take out your materials, and listen to the conversation.

"We've invited you to our meeting today," Robert begins, "because we need a marketing letter written on our low-wattage electrical sidewalk system known as *Hot Blocks*. We're pretty excited about the product, especially for cold weather cities, and now that our testing phase is just about completed, we need the letter written."

"Okay," you respond. "I think I'll just absorb some of the conversation, and then ask a few questions. Will I have access to your notes and lab reports?"

"We'll give you copies of whatever we think you need to write the letter," Robert quickly responds. You detect a slight edge in his voice but decide you are being a little suspicious.

Robert addresses the group, "We've completed the preliminary testing phase on *Hot Blocks,* and we're pretty confident that this product is ready to go. In fact, the administration at Eastern Wisconsin University is ready to sign a pretty lucrative contract with us to install the sidewalks throughout the university." Robert looks at you, "That's why we need you to write up the letter about the product's features and benefits so quickly."

"Why a regional university and not a major university?" another engineer, Luther Blackwell, asks. "Usually the big schools have the money for something like this."

"Well, apparently a student was badly injured at Eastern last year," Cruzner continues. "She slipped on the ice as she entered the university's administrative center, fracturing her hipbone and breaking her hand. There's been talk of a lawsuit. I guess Eastern feels that it's demonstrating goodwill to install a system like this—before the case is filed."

"Are they ready to commit the finances?" Janice Blake, a systems engineer interjects.

"Not until they receive our report indicating that the product is safe, reliable, and ready to go," Cruzner adds. "I think we're at that point, but I want to go over the testing with you one more time."

For the duration of the meeting—about an hour and a half—Robert Cruzner evaluates all aspects of testing the *Hot Blocks* product. You take notes, writing down key figures and adding questions you want to ask later. You notice that most of the testing on the product has been in lab-approximated temperatures of freezing or slightly below zero. Nothing under 10 degrees below zero is mentioned. You need to ask Cruzner about that later. The meeting breaks up just before lunchtime, leaving you time to ask Cruzner a few questions before you retrieve your tuna sandwich out of the lunchroom.

"Robert, I have a couple of quick questions," you say as he's about to leave the room. "Can you stay just a moment or two longer?"

"Sure," he replies. "Let's do it now because I have meetings on this all afternoon."

"Okay," you say, "this will be quick." "The *Hot Blocks* product is manufactured here on the premises, right?"

"Yes."

"And the product is a premade concrete plate with low wattage circuitry running underneath the plate that then mates with the existing sidewalk?"

"Well, yes. It's a little more complicated than that, but for your purposes that's adequate."

You detect a little condescension in his voice, but you ignore it. "In the notes I took this morning, I only heard mention of the lowest temperature testing being something like ten below. Is that right?"

Cruzner pauses before speaking. "That's low enough for our purposes and the contract with Eastern," he says. "Write the marketing materials based on the comments this morning and the notes I will give you. You realize, of course, that I reserve the right to read your draft before it goes to Eastern."

Actually, you weren't aware of that. Your immediate supervisor is Jason Monroe, the Products Manager. To your knowledge you reported to no one else. "I'll give the marketing letter to Jason first, Robert," you reply. "If you want to review it, he'll have it."

Cruzner pauses and looks at you. The beginnings of a smile develop around the corners of his mouth. "That will be fine," he finally concludes. He turns and walks away.

You leave the conference room and head back to your office to begin logging your comments into the computer. You read through the materials that Cruzner and his team of engineers provided you, make notes on potential marketing aspects of the new product, and arrange your ideas in a rough outline. You look forward to drafting this marketing letter, the first of your professional career.

After lunch, you return to your office to begin drafting the report. On the top of your desk is an unmarked, unsealed envelope. You turn over the envelope and pull at the flap tucked inside the envelope. Inside you find two sheets of paper, photocopies of something, and nothing else—no note, no indication of who sent it, nothing. Your first temptation is to go straight to Jason, but you decide to settle down and read whatever's on the two pages before making any snap judgments.

The first sheet is a photocopy of a memo from Luther Blackwell to Robert Cruzner. You are surprised to see that it is dated February 18, 1996 —a month ago. You read the memo.

MEMORANDUM

TO: Robert Cruzner

FROM: Luther Blackwell

DATE: February 18, 1996

SUBJECT: *Hot Blocks*

Bob, I've had a chance to do that testing you asked me to do, so I'm attaching to this memo my notes. I tested the *Hot Blocks* product at temperatures ranging from 32 degrees Fahrenheit all the way down to 50 degrees below zero (Fahrenheit). You wanted me to test the run-off of melting snow according to temperature, and I see no problems in rapid runoff or accelerated melting. The overall system looks good.

At 50 below, complete system integrity cannot be guaranteed, but such temps are rare and pose no problem at this time.

Give me a call if you have further questions.

The second sheet of paper, stapled to the memo, is a photocopy of what you presume to be Luther Blackwell's notes:

Reports of low wattage circuitry and heated sidewalks (known as "Hot Blocks")
Luther Blackwell, Test Engineer

I like the durability of our product @ the freezing point. Snow of approx. 6" melts @ rate of 1/2" per hour w/ no discernible puddling of runoff due to rapid melting. Looks good to me.

Rate distribution chart below—
6" (32° F) 1/2" runoff per hour
6" (20) 1/3" rph
6" (0) 1/4" rph
6" (-20) 1/8" rph
6" (-30) 1/16" rph
6" (-50) and below less than 1/16" runoff per hour, but rare conditions. (Few hairlines @ circuitry base, but minor and temps rare. Can't guarantee won't be cracks past a certain point.)

You finish reading the memo and the notes and you are amazed first of all that someone unknown to you "happened" to leave them on your desk. You are also amazed to discover that in fact temperature testing had been done at temperatures well below minus ten and with mixed results. You are well aware of the weather in that region and a fifty-below day is not out of the question. The new material raises many new questions as you prepare to draft this letter that is supposed to sell Eastern on the product. You have to think through carefully your response to this problem.

… Scenarios and Case Studies   125

Case Study 3: Making Waves

Making Waves

CONSIDERATIONS

You are a part-time employee seeking to distinguish yourself in the kind of job you've always wanted. When asked to assist your boss, you jump at the chance. While reading through this situation, keep the following questions in mind:

- What technical skill levels will the document's readers posses?
- What are the problems associated with the wave-making device?
- How do these problems affect the creation of your document?

It took you two years, but you finally landed a summer job at the New City Aquarium. Your main responsibilities are helping to feed and clean up after the sea lions and otters and any other tasks required by your supervisors. While it has nothing to do with your studies in engineering, you enjoy working with the animals and the mental break the job provides.

Today, you report to work and are surprised to find a note taped to your locker:

> Please report to my office as soon as possible. We have an important project that requires your help.
>
> —J. Miller

The note's author is Jim Miller, a manager in the aquarium development projects division. You only met Jim once at a barbecue to welcome new staffers. You can only remember Jim as a very stocky man who looked more like a fisherman than a man who researches, plans, and helps fund new exhibits.

Jim greets you at his office door with a hearty handshake.

"What can I do for you?" you ask.

"I'm sure you're surprised to be called up here," Jim says. "But I remember you mentioning that you were studying engineering, and I have a project requiring that kind of work."

"I'm not an engineer yet," you explain.

"I know," Jim smiles. "I just need some help untangling what a real engineer worked up for our new shoreline birds exhibit."

"What's the problem?" you ask.

"It's these plans for the wave-making machine," Jim explains, handing you a manila folder. "I need to explain how it works for this grant I'm writing, and I 'm not sure I can figure it out myself."

"We're getting a wave-making machine?" you ask incredulously. New City's aquarium is relatively small, with a proportional budget. "Why do shoreline birds need waves?"

"Ah, engineers with no understanding of ecology," Jim jibes, but with good humor. "Shoreline birds feed off the small creatures that wash up on the beach. That's why it's so important to preserve specific types of beaches with the right mix of sand and rock—and the right wave height. We want to do more than just dump food at these birds. We need to demonstrate their dependency on the beach itself and the waves that feed them."

You nod appreciatively. "I have to explain how this machine is unique," Jim continues. "We've got to have a special kind of wave—not too big, not too small, and at the right intervals. And it can't make a smacking noise—that disturbs the birds. We had an engineer friend of mine develop it, but his specifications read like Greek to me. I need a complete description of the machine, but one that the people reading the grant will understand."

"I can work up a technical description for you," you say. "It may take me a little while, though."

"We have about a week," Jim says. "And I'll need you to do this in addition to your normal work. You'll get overtime pay for this project."

Your smile widens, and Jim laughs. "Get me a draft as soon as possible," he says and turns back to the work on his desk. You leave Jim's office, stuff the folder in your locker, and change quickly. The otters will be hungry soon.

At home, you open the folder and study its contents. You find a drawing of the wave machine (Figure 1).

From what you can tell, the spindle arm rotates the oblong-shaped drum at a constant speed; however, the shape of the drum causes water to be displaced unevenly, creating the wave. The water that bounces back from the shore crests on the drum and is sent back as a smaller, gentler wave. The en-

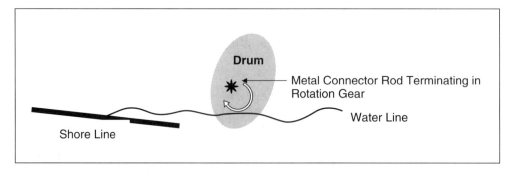

FIGURE 1   The Wave Machine

gineer estimated that five measurable waves occur with each rotation, with the first wave the largest and the rest decreasing in strength. A computer controls the rotation speed and thus the size and frequency of the waves. You see a copy of the engineer's report, which includes the following information:

> Mechanism produces volume distortion to the amount of 15 inches up a 10-degree "shore" slope. Succeeding distortions measured 9, 6, and 4 inches up the 10-degree slope. Some counter-wave action occurs every fourth rotation, with volume displacement measuring approximately 7 inches, with a 3, 2, and no measurable displacement until the next rotation. Some significant disturbance of sand under the drum occurred. Recommend using gravel under the drum and approximately 5 feet in front of and behind—should prevent excessive clouding of the water. Operation of the drum did not affect feed species population until water visibility dropped to near zero. NOTE: Heat from drum operation resulted in algae growth of unacceptable species (as determined by J. Miller). Problem not yet resolved; however, several low-heat prototypes are in the design phase.

A few days later, you catch up with Jim to ask him a few questions. "How much do you want me to include? I noticed that the engineer is having a problem with algae."

"Yeah," Jim says, "but he's close to solving that one, and we'll have it taken care of before we actually build the machine. Right now, we just need the money to start it up. The funders need to have confidence in this project."

You think you understand Jim's intent, but before you have a chance to talk further, he explains he is late for a meeting, asks you for a draft by tomorrow, and leaves you standing in the hallway with a sheaf of papers and unanswered questions. You realize you have no choice but to get a draft to him by the next day.

Case Study 4: Where's the On-Ramp to the Information Superhighway?

Where's the On-Ramp to the Information Superhighway?

CONSIDERATIONS

As someone who is learning to use new technologies as part of your college studies, you are often the perfect candidate for teaching your own new skills to others. In this situation, you are also saving yourself and others time by not repeating information. As you read the scenario, keep the following questions in mind:

- Think back to the first time you used the Internet, perhaps at your school's computer lab. What were your questions, frustrations, or mistakes during that experience?
- What is the skill level of your readers?
- What considerations, other than the Internet log-in procedures, are involved?

You are just starting your sophomore year at Technical University, a small four-year college in Hampton, Idaho. During your first year, you worked as a dishwasher in the University Center's kitchen. While the money kept you afloat during the school year, it wasn't the best job you'd ever held. This year, you have landed a position as a tutor in the campus computer lab.

As a tutor, you work in the lab, roaming between stations and helping students with their computing problems. Most of the time, you answer the same questions over and over, and sometimes you have to tell the same students the same thing several times. While you find this frustrating, you have not yet lost your patience, but you wonder how long you can remain completely polite.

The most common question you are asked concerns the procedure for logging on to the Internet and checking e-mail. The computer labs recently changed how students access their e-mail accounts because of security problems. You decide to talk to the computer lab supervisor, Melissa Swadley, about creating some kind of handout to give students seeking to surf the Net.

"That's a good idea," Melissa says. "We've been meaning to get something in writing, but no one has the time. If you want to tackle this, go ahead."

"Great," you say. "How much should I explain? Just the Internet or e-mail or both?"

"Both, definitely," Melissa says. "Maybe you can fit the instructions on the same page. You know, Internet on one side and e-mail on the other."

"Just one page?" you ask, thinking that the procedure is a bit complex, especially if you are going to include troubleshooting information.

"Yeah, keep it short and simple," Melissa says. "Besides, we don't have a big budget and I want to be able to photocopy this off whenever we need it. I don't have the money to take this to the campus print shop."

"I'll give it a try," you say. You sit down at a station and begin to work through the e-mail process, taking notes as you go. *This isn't all that simple,* you think, remembering how computer illiterate some of the students you have helped were.

Case Study 5: Usage as an Interactive Strategy for International Team-Building

Usage as an Interactive Strategy for International Team-Building

The Never-Ending Story

BOYD H. DAVIS
University of North Carolina at Charlotte

JEUTONNE BREWER
University of North Carolina at Greensboro

YE-LING CHANG
National Kaohsiung Normal University

Abstract: In this case, a technical communicator discovers that even the "little touches"—such as pronouns—are important in cross-cultural collaboration. Because his business is getting ready to train employees in different countries, the technical communicator must harmonize the training scripts to achieve consistency. He finds that, since different languages handle gender reference differently, translators and writers at each site will need to cooperate and share insights. The case asks readers to consider the difference between grammar and usage, and to discuss such questions as: How can technical communicators work with the notion of "face" in cross-cultural situations? What does a technical communicator need to know about English usage in general, and issues of nonsexist usage in particular, when establishing strategies for cooperation among international writers?

Note: To the best of our knowledge, there is no American pizza-restaurant chain in Asia or the United States precisely like the one we describe. However, the issue that Alex faces—harmonizing document design across cultures by paying attention to details that *at first seem to be minute, even trivial—is not going to vanish. The information in and the wording of the faxes and e-mail messages in the case discussion are adapted from an interactive e-mail exchange among writers in Taiwan, Japan, and North Carolina developed and moderated by the authors in 1995.*

Background for the Case

Pizza-for-Us is a franchised chain that started in Tucson a decade ago and spread rapidly to both coasts. While relatively young in the fast-food business, it's considered a solid investment on Wall Street, and buying a franchise is worth the high price. Management in the home office astutely profiles a variety of consumer habits in a particular area before selling a franchise. The home office closely monitors performance and quality and is adamant that the "special experience" at Pizza-for-Us restaurants must remain absolutely consistent.

While Pizza-for-Us will deliver pizza to consumers, its emphasis—and its high-profile advertising thrust—is on its family atmosphere. Each franchise site presents a cozy setting and a specific layout. There are no booths; instead, the restaurant has circular tables with a lazy Susan in the middle of each one. Family-sized, topping-rich pizzas are served on the roundtables. Customers share a pizza and rotate the lazy Susan to sample the variety of toppings on each eighth of the pizza. These toppings, like the decorative touches to the dining area, are regionally cued whenever possible: Pizza-for-Us keeps field managers busy charting regional demographics so it can move swiftly to target various niches. Service is fast, polite, and exceptionally well-rehearsed, with a longer-than-usual training period. Other restaurants eagerly recruit workers from Pizza-for-Us, from the people clearing the tables through the servers and on to the host or hostess.

For the last two years, field inspectors have noted that the restaurants in areas with large Asian populations consistently get repeat business; more patrons return to Pizza-for-Us restaurants than to other restaurants in the same price category. That has added to management's confidence; the chain is now ready to open franchises in Taipei, Tokyo, and Seoul, even though a number of other U.S. fast-food restaurants have already established a presence there. Negotiations have been successful; the local agreements are in place; construction is almost finished; and the training process for employees is about to start. So is Alex's problem.

Introducing Alex

Alex Calkins is a junior technical writer in the Publications Department at the Pizza-for-Us corporate headquarters in Tucson. His college internship was with the information services division of a software company, and he worked after graduation on a contract to develop manuals for a utility company, but this is his first real career position. Since he is new, his manager, Christa Johnston, has given him responsibility to develop only one part of the training materials for the waitstaff and has already cautioned him that at Pizza-for-Us, consistency in details is taken seriously.

Usage as an Interactive Strategy for International Team-Building   **3**

"What you'll need to do for next week," she says, "is work on these training scripts. These don't have the special jargon that we'll teach the servers to use with the kitchen staff. You can look at that later. Right now, go over the way the host greets and seats the customers and how the servers highlight the menu and take orders. The writers at our new sites are busy translating the scripts. You'll handle their questions; some came this morning."

Later that day, Christa comes back to Alex with a thin manila folder. She's really in a hurry, he thinks. "You're going to have to do this by Friday," she says.

Alex's Problem

"Friday?" asks Alex. This is Monday, and the project so far seems pretty straightforward. "Oh, right," says Christa. "We think it would be a good idea if you brought your recommendations to staff meeting. We're going to try to streamline the whole training manual." She hustles down the hall. Back at his desk, Alex opens the folder. Clipped to the script are printouts of e-mail messages to Christa.

On top is one from Seoul. Mr. Lee wants advice on translating some of the sentences describing the menu and phrases for greetings. They don't match his expectations for correct English. The greetings seem a little informal to him. And what's behind the use of "s/he" in the script for the hostess: "S/he will be with you in just a moment to take your order." Is there some special reason for this usage?

Next is the printout of an e-mail message from Tokyo sent the same day. It's about a new issue—or is it? Mrs. Hatayama is concerned about the word *hostess*: Is this like a receptionist? She wants to offer a slightly different translation, since older Japanese might expect a cabaret hostess. And would he advise her on the use of *it* in phrases like "It's been great to serve you"? Japanese don't use an equivalent for this pronoun when *it* is a "filler," so can she change that a bit?

Alex pauses: *It* as a "filler"? He looks at Mr. Chen's e-mail from Taipei. Mr. Chen is concerned with the bulleted directions for the server. Those bullets could translate as commands. Could he soften the tone a bit to make sure the directions were requests? And is there a reason for using "he or she" to refer to the servers? What if they are all the same gender? Can't they just delete the pronoun in the English version, just like the Chinese? Customers will understand, and so will the staff.

Alex puts down the folder. He goes to see Lita.

Lita's Response

Lita Ganzer is in the training division, where she offers seminars on different aspects of communications. Alex had taken her short course on avoiding sexist usage and has signed up for the training series on cross-cultural communications—but it starts after Friday's meeting.

"You're right," says Lita. "This is a cross-cultural issue—in fact, there are several issues. The writers of our training scripts have been very careful to avoid what Americans consider sexist usage in English. Each of your translators is having problems with the script—that may be because of differences between their lan-

4 Usage as an Interactive Strategy for International Team-Building

guages and English. But there's another issue here as well. I think there are probably some differences in cultural expectations. I'm looking at that in the seminar next month. Did you sign up for it?"

"My meeting is this Friday," said Alex.

"I have a meeting in a few minutes," said Lita. "But I'll lend you my first set of slides for that workshop. Take a look at them—you can open them on your computer—and see if they make sense to you. I could see you later, around 3:30, if you want to talk this through. I've got a strategy for team-building that you might want to use."

"I'll see you then and bring you some coffee," replies Alex. He takes the diskette from Lita and goes back to his desk.

Lita's Slides

Alex studies the slides (See Figure 1). How, he thinks, is he going to pull everyone together? He doesn't have any idea of what the writers in Taipei, Tokyo, or Seoul might think is culturally appropriate.

FIGURE 1 Involvement and Independent Strategies

Cross-cultural teamwork builds on cross-cultural communication. Use these insights taken from Scollon and Scollon (1995) to understand the communication situations illustrated in this case

When people communicate with each other, part of their message signals the degree to which they see themselves as being	• involved with each other • independent from each other
Communicators use various strategies to signal something about	• their own identity • their attitude toward the communication situation • how they want to preserve face in the situation
Involvement strategies signal alignment with the other person by claiming	• a common point of view • in-group membership *Example: All of us at PFU hope you'll join us in thinking this training process is the best we've ever developed. We keep your life in mind.*
Independent strategies signal some sort of distancing from the other person through an awareness of different	• points of view • group affiliations *Example: This notifies you of company policy about . . .*

Both strategies (involvement and independent) are "correct," but different groups and different cultures have different notions about which strategy is appropriate.

Usage as an Interactive Strategy for International Team-Building 5

"That's actually going to be a strength," says Lita when he meets with her later that day. "You know that you don't know something, and you don't have any problem admitting it. That makes you ready to learn from other people. Neither you nor the other writers can be totally sure of what's appropriate in each culture or each language. You're going to have to think of a way for all four of you to be involved with each other in a common task, where each person has something unique to contribute. I have an idea for you."

"I'll take it," says Alex.

"Pronouns," said Lisa. "It's pretty obvious from the messages you shared with me that all of them are having some problems with how to translate the pronouns. English pronouns signal gender, which is why I went over pronouns in the non-sexist usage workshop. There are cultural expectations about gender roles, and pronouns can trigger those expectations or assumptions. Remember my examples about managers? They're certainly not all 'he's' and we don't want to suggest that they are."

"What about other languages?" asks Alex. "I speak Spanish, and it has gender."

"Well," said Lita, "I know that Mandarin doesn't. Here's your chance to find out about the others, and you can go from there to see what kinds of cultural associations are triggered. Here's why: All of your team members have unique information about language use and culturally cued expectations at their home site, just like you do. And all of you have access to different kinds of authority, like textbooks, grammar books, and dictionaries. These books prescribe correctness. Usage manuals, guidelines, or stylesheets let you know what's generally appropriate. If you can specify a focus, and position yourself and the other writers as equal experts, what you as writers report about your own practices will give you even more information."

"Do you mean," asks Alex, "if we all share information about how we write or what we have problems with, we can get a handle on cultural issues, too?"

"I like your use of 'we,'" smiles Lita. "I'll look at your plan if you want."

Alex goes back to his office to write a message. Later, he shows it to Lita, and after revising his phrasal verbs and colloquial expressions, he sends it to the other writers. He is already thinking of the writers as "our team."

Alex's E-mail: His Team-Building Strategy

On Tuesday morning, Alex logs on and finds messages waiting (See Figure 2). They're already deep in discussion: How can this be? He kicks himself mentally for forgetting the time difference. They're already through with Tuesday. They must have found the message first thing in the morning, and have gotten busy right away. He is lucky to be working with them, he thinks. He prints out the first set of messages and lays them out in chronological order. (See Figures 3, 4, and 5).

Quickly, Alex writes up the results of his own office survey, and sends it on. (See Figure 6).

The next morning, Alex finds another group of messages (See Figures 7, 8, and 9).

6   *Usage as an Interactive Strategy for International Team-Building*

> **To:** Mr. Chen, Taipei Office, Pizza-for-Us
> Mrs. Hatayama, Tokyo Office, Pizza-for-Us
> Mr. Lee, Seoul Office, Pizza-for-Us
>
> **From:** Alexander Calkins, Tucson Office, Pizza-for-Us
>
> **Re:** Training Scripts
>
> Dear Colleagues,
>
> Our manager, Christa Johnston, has asked me to work with you on the training scripts. All of us share a concern for quality and a concern that the language used in the training scripts will be appropriate. Your comments and questions are very useful, and I will take them to the Editing Team meeting on Friday to discuss how best to revise the training scripts so they can be suitable to each of your locations. I see that all of us have concerns about translating English pronouns. So that I can better understand these concerns, could you please e-mail me by Thursday evening (Tucson-time) some additional information in response to my questions for you about pronouns?
>
> In English, as you know, pronouns signal gender, which can spill over into ideas about cultural issues such as jobs or even ways to greet people or refer to them. All of those issues are important if the training scripts are to be consistent and also appropriate. Here is a list of sentences, each of which might present a different kind of problem. Could you please choose several of these, and informally ask several people for their responses? (Would they be more likely to use "he" or "she"?) I would greatly appreciate your assistance.
>
> 1. The professor/The teacher won't dismiss our class early, will ___?
> 2. The chef/The cook was congratulated by the guests, wasn't ___?
> 3. Either Ian or Yoshi/Maria or Keisha will stay home, ___?
> 4. Germany/America/The U.S. supports the U.N., ___?
> 5. The interior decorator/lighting designer will finish soon, ___?
> 6. My dog/my cat/my parakeet will eat anything, ___?
> 7. Neither Yusuf nor Willem stayed at the office, ___?

FIGURE 2 Sharing Concerns

Usage as an Interactive Strategy for International Team-Building   7

To: Alexander Calkins, Technical Writing Team

From: Mrs. Hatayama, Tokyo Office, Pizza-for-Us

Re: Pronouns

On #1: I got "does she" from one man and "does he" from the women and from one other man. This shows that the word /teacher/ gives us the image of he. Compared to this, /interior decorator/ seems to have the image of a "she." I thought it would be a "he," though. Also, what about using "they" when the subject looks singular? I hear this in the movies.

FIGURE 3   **Discussing Pronoun Usage**

To: Alexander Calkins, Technical Writing Team

From: Mr. Chen, Taipei Office, Pizza-for-Us

Re: Pronouns [in # 1 and #5]

Problems with #1 and #5. Mandarin has a very different system for pronouns. The "we" of speaker-hearer is indicated by a pronoun different from the pronoun for speaker-hearer-other person. This is very different for English. When we studied English, our professor said that learning the chart for pronouns wasn't enough, we had to look at situations for their use.

FIGURE 4   **More Discussion of Pronoun Usage**

To: Alexander Calkins, Technical Writing Team

From: Mr. Lee, Seoul Office, Pizza-for-Us

Re: #4, pronouns and countries

Our reference books say to use the singular, but the people in my office don't do that for either the U.S. or North America. Also, we have some British speakers on our staff, and when I asked them the questions, they inserted "Right?" instead of the pronoun. Is this acceptable? Also, is it conventional now to use "they" where my books say "he"?

FIGURE 5   **Pronouns and Countries**

8   *Usage as an Interactive Strategy for International Team-Building*

> **To:**   Mr. Chen, Taipei Office, Pizza-for-Us
> Mrs. Hatayama, Tokyo Office, Pizza-for-Us
> Mr. Lee, Seoul Office, Pizza-for-Us
>
> **From:**   Alexander Calkins, Technical Writing Team
>
> **Re:**   "They" for countries and groups of people
>
> For the U.S., half of the people I asked use "it" and a fourth use "they." The others use "he" and "she". Using "they" is considered colloquial and sometimes, nonstandard. Many people use "they" in speech, but usually not in formal written texts, even though famous authors use it. This pronoun was a topic in our Nonsexist Usage workshop. I'm mailing you notes from that workshop so you can see some of our language constraints.

FIGURE 6   Pronouns Continue

> **To:**   Alexander Calkins, Technical Writing Team
>
> **From:**   Mr. Chen, Taipei Office, Pizza-for-Us
>
> **Re:**   Animals
>
> This could be important for ads and layout. People in my office say "it" for dogs and "she" for cats, sometimes for birds, when speaking English. Also, what about colors? Any gender associations there besides for babies' clothes?

FIGURE 7   Referring to Animals

> **To:**   Alexander Calkins, Technical Writing Team
>
> **From:**   Mrs. Hatayama, Tokyo Office, Pizza-for-Us
>
> **Re:**   Colors and Animals
>
> We have "he" for dogs and birds and "it" for cats. But we don't specify the pronoun unless we're talking in another language. Purple is imperial, white is for funerals, red is for festivals. I think we differ from American usage.

FIGURE 8   Colors and Animals

Usage as an Interactive Strategy for International Team-Building 9

> **To:** Alexander Calkins, Tucson Office, Pizza-for-Us
>
> **From:** Mr. Lee, Seoul Office, Pizza-for-Us
>
> **Re:** Changing the Subject
>
> If you don't mind, I want to ask about servers and how they greet. When I was in Tokyo last spring, I went to an American hamburger restaurant. All the servers greeted me in a chorus. That's not in our training script, and it could make problems.

FIGURE 9   Servers and Greetings

Scenario 1: Writing a Proposal

> Imagine you have just been assigned to an advisory committee that has been asked to develop ways to reduce the number of cars being driven to your campus or workplace. What information do you have already on this topic? What information would you still need to solve this problem? What are some questions you would need answered before you and your team began writing a proposal to solve the problem?

Scenario 2: Writing a Proposal

> You have been asked to develop a mentoring program at your college or workplace. The current problem is that new students or employees often feel overwhelmed by the immediate onslaught of work. As a result, they often drop out or quit within a couple of months. Your task is to set some goals, map out a solution, and write up a two-page description of your mentoring plan. Your plan should answer the how and why questions while providing some tangible deliverables.

Sources

Bosley, Deborah S. *Global Contexts* © 2001. Reprinted by permission of Pearson Education, Inc. Case Studies 1 and 5.

Gurak, Laura J and John M Lannon, *A Concise Guide to Technical Communications* © 2000. Reprinted by permission of Pearson Education, Inc. Proposal 2.

Gurak, Laura J., *Oral Presentations* © 2000. Reprinted by permission of Pearson Education, Inc. PowerPoint 1.

Johnson-Sheehan, Richard, *Technical Communication Today* © 2005. Reprinted by permission of Pearson Education, Inc. Letters 3 and 4; Memos 3 and 8; Career Letter 3, Résumé 3; Report 5.

Johnson-Sheehan, Richard and Sam Dragga, *Writing Proposals* © 2002. Reprinted by permission of Pearson Education, Inc. Scenarios 1 and 2.

Jones, Dan, *The Technical Communicator's Handbook* © 2000. Reprinted by permission of Pearson Education, Inc. Letter 8; Career Letter 5; and Résumé 2.

Jones, Dan and Karen Lane, *Technical Communication: Strategies for College and the Workplace* © 2000. Reprinted by permission of Pearson Education, Inc. Description 3, Report 2, Abstract 1.

Keller, Arnold, *The Practical Technical Writer* © 2004. Reprinted by permission of Pearson Education, Inc. Proposal 1.

Kynell, Teresa C. and Wendy Kreig Stone, *Scenarios for Technical Communication: Critical Thinking and Writing* © 1999. Reprinted by permission of Pearson Education, Inc. Letters 2, 5, and 11; Memo 4; Case Studies 2, 3 and 4.

Lannon, John M., *Technical Communication*, 10th Edition © 2006. Reprinted by permission of Pearson Education, Inc. Memo 7; Career Letter 4; and Proposal 4.

McMurrey, David, *Online Technical Writing: Online Textbook* © 2000. Reprinted with adaptations by permission of David McMurrey. Proposal 3.

Moore, John H. and Christopher C. Davis, *Building Scientific Apparatus* © 2002. Reprinted by permission of Perseus Books Publishers, a member of Perseus Books, L.L.C. Description 2.

Pearsall, Thomas E., *The Elements of Technical Writing, Second Edition* © 2001. Reprinted by permission of Pearson Education, Inc. Career Letter 3, Reports 1, 3 and 4.

Pickett, Nell Ann, Ann Laster Appleton, and Katherine E. Staples, *Technical English: Writing, Reading and Speaking*, 8th Edition © 2001. Reprinted by permission of Pearson Education, Inc. Letter 7; Memo 4; Résumé 1; and Instructions 2.

Reep, Diana C., *Technical Writing: Principles, Strategies, and Readings, Fifth Edition* © 2003. Reprinted by permission of Pearson Education, Inc. Career Letter 1.

Searles, George J., *Workplace Communications: The Basics, Second Edition* © 2003. Reprinted by permission of Pearson Education, Inc. Letters 1, 10, and 12; Memo 5.

Woolever, *Writing for the Technical Professions,* 2nd Edition © 2002. Reprinted by permission of Pearson Education, Inc. Letters 6 and 9; Instructions 1; and Description 1.

NOTES

NOTES

NOTES

NOTES

NOTES

NOTES

NOTES

NOTES

NOTES

NOTES

NOTES

NOTES